CAMBRIDGE LIBRARY COLLECTION

Books of enduring scholarly value

Technology

The focus of this series is engineering, broadly construed. It covers technological innovation from a range of periods and cultures, but centres on the technological achievements of the industrial era in the West, particularly in the nineteenth century, as understood by their contemporaries. Infrastructure is one major focus, covering the building of railways and canals, bridges and tunnels, land drainage, the laying of submarine cables, and the construction of docks and lighthouses. Other key topics include developments in industrial and manufacturing fields such as mining technology, the production of iron and steel, the use of steam power, and chemical processes such as photography and textile dyes.

The Chemistry of Light and Photography

First published in 1875, this book describes the history and science of photography, with an emphasis on the practical processes involved and their relation to the physical and chemical properties of light. Hermann Vogel (1834–1898), who pioneered the technology for colour photography, was Professor of Photography at the Royal Industrial Academy of Berlin. Here he explains the science of photography simply and concisely for a popular readership. The book has 100 illustrations, including both woodcuts and 'specimens of the latest discoveries in photography', intended to demonstrate the value of the technology to society. It traces the history of photography from its beginnings in experiments conducted by Davy and Wedgwood and the invention of the Daguerreotype to the most recent developments of Vogel's day. It was regarded as the most comprehensive guide to photography then available, and ran to several editions. This reissue is of the fourth edition of 1883.

T0280212

Cambridge University Press has long been a pioneer in the reissuing of out-of-print titles from its own backlist, producing digital reprints of books that are still sought after by scholars and students but could not be reprinted economically using traditional technology. The Cambridge Library Collection extends this activity to a wider range of books which are still of importance to researchers and professionals, either for the source material they contain, or as landmarks in the history of their academic discipline.

Drawing from the world-renowned collections in the Cambridge University Library, and guided by the advice of experts in each subject area, Cambridge University Press is using state-of-the-art scanning machines in its own Printing House to capture the content of each book selected for inclusion. The files are processed to give a consistently clear, crisp image, and the books finished to the high quality standard for which the Press is recognised around the world. The latest print-on-demand technology ensures that the books will remain available indefinitely, and that orders for single or multiple copies can quickly be supplied.

The Cambridge Library Collection will bring back to life books of enduring scholarly value (including out-of-copyright works originally issued by other publishers) across a wide range of disciplines in the humanities and social sciences and in science and technology.

The Chemistry of
Light and Photography

In their Application to
Art, Science, and Industry

H ERMANN V OGEL

CAMBRIDGE
UNIVERSITY PRESS

CAMBRIDGE UNIVERSITY PRESS

Cambridge, New York, Melbourne, Madrid, Cape Town, Singapore,
São Paolo, Delhi, Dubai, Tokyo, Mexico City

Published in the United States of America by Cambridge University Press, New York

www.cambridge.org
Information on this title: www.cambridge.org/9781108026628

© in this compilation Cambridge University Press 2011

This edition first published 1883
This digitally printed version 2011

ISBN 978-1-108-02662-8 Paperback

THE INTERNATIONAL SCIENTIFIC SERIES.
VOL. XV.

PLATE I.

PHOTOGRAPH OF THE MOON,
(From RUTHERFURD's original negative.)

See p. 188.

THE CHEMISTRY OF LIGHT

AND PHOTOGRAPHY

IN THEIR APPLICATION

TO ART, SCIENCE, AND INDUSTRY.

BY

DR. HERMANN VOGEL,

PROFESSOR IN THE ROYAL INDUSTRIAL ACADEMY OF BERLIN.

WITH 100 ILLUSTRATIONS.

FOURTH EDITION.

LONDON:

KEGAN PAUL, TRENCH & CO., 1, PATERNOSTER SQUARE.

1883.

PREFACE.

———◦◦◦———

AMONG the splendid scientific discoveries of this century, two are specially prominent—Photography and Spectrum Analysis. Both belong to the province of Optics, and at the same time to that of Chemistry. While Spectrum Analysis has, down to the present time, remained almost exclusively in the hands of scientific men as a most important means of research, Photography passed immediately into practical life, spread over almost every branch of human effort and knowledge, and now there is scarcely a single field in the universe of visible phenomena where its productive influence is not felt.

It brings us faithful pictures of remote regions, of strange forms of stratification, of fauna, and of flora; it fixes the transient appearances of solar eclipses; it is of great utility to the astronomer and geographer; it registers the movements of the barometer and thermometer; it has become connected with porcelain painting and the various forms of printing; it makes the noblest works of art accessible to those of slender means.

But it does more than this. A new science has been called into being by Photography, the Chemistry of Light; it has given new conclusions respecting the operations of the vibrating ether of light. It is true that these services, rendered by Photography to art and science, are only appreciated by the few. Men of science have in great measure neglected this subject after the first enthusiasm excited by Daguerre's discovery had passed away; but seldom is Photography mentioned in the text books of physics and chemistry.

This consideration has induced the **Author** to lay before the public a popular treatise of Photography and the Chemistry of Light, showing their important bearing on science, art, and industry. The Publisher has met the Author in the readiest manner, not only by providing numerous woodcuts to explain the text, but by obtaining specimens of the latest discoveries in Photography, at a considerable cost; so that the illustrations afford the reader an opportunity of seeing of what value Photography can be in connection with the printing press. The Author trusts the work may meet a friendly reception.

THE AUTHOR.

Berlin, *January*, 1874.

TABLE OF CONTENTS.

———◆◇◆———

THE CHEMISTRY OF LIGHT.

CHAPTER I.

DEVELOPMENT OF OUR PHOTO-CHEMICAL KNOWLEDGE.

Various Effects of Light—Physical and Chemical Changes—Bleaching
Effect of Light—Action of Light upon Chloride of Silver and Lunar
Caustic (Nitrate of Silver)—Chemical Ink—Pictures upon Paper
prepared with Nitrate of Silver—The Researches of Wedgwood and
Davy — The Camera Obscura — Nièpce — Effects of Light upon
Asphalt—Heliography—Its Application to the Production of Paper
Money—Iodide of Silver—Discovery of the Daguerreotype.

THE light which radiates from the great central body of
our planetary system produces manifold effects upon the
animate and inanimate world, some of which are at once
evident to the senses, and have been known for thousands
of years, while others, again, are not so apparent to the
eye, and have been discovered, examined, and utilized
only through the observations of modern times.

The first effect which every person, however unculti-
vated, notices when, after the darkness of night, the sun

rises, is that bodies become visible. The rays from the source of light are thrown back (reflected) from various bodies, they reach our eyes, and produce an impression upon the retina, the result of which is the perception of material objects by the eye. But soon another effect is observed, not by the eye, but by the sense of feeling. The sun's rays not only illumine bodies upon which they fall, but heat them, as is felt when the hand is held in the rays. Both effects, the shining, or illuminating, and the warming effects of the sunbeams, differ very essentially from each other. The illuminating effect we perceive instantaneously; the heating effect is only felt after a certain time, which may be shorter or longer, as the heating power of the sun is stronger or weaker.

In addition to these two effects of sunlight, there is a third, which generally requires a still longer period to make itself noticed, and which cannot be directly perceived either by the eye or by the sense of feeling, but only by the peculiar changes which light produces in the material world. These are the chemical effects of light.

If we take a piece of wood, and bend or saw it, we change its form; if we rub it, it becomes warm—we change its temperature, but it still remains wood. These changes, which do not affect the substance or matter (*Stoff*) of a body, we term physical.

But if we set fire to a piece of wood, strong-smelling gases ascend, ashes are deposited, and a black residue remains, which is totally different from the wood. By this process a new substance—charcoal—has been produced. *Material* changes of this kind we term chemical changes;—and such chemical changes are, in an especial manner, produced by heat. If, for instance, we heat a bright iron wire red hot, it undergoes apparently only a physical (not a material) change. But, if we allow it to cool, we find the bright rod has become dull and black; that it has acquired a brittle, black surface, which easily breaks away on bending the rod, and differs entirely from the bright, tough, flexible iron. Here again a chemical change, that is to say, a change of substance, has taken place; the iron has been converted into another body, into iron scale, because it has combined with a component part of the surrounding air—with oxygen.

Chemical changes of this kind are not only produced by heat but also by light.

It has long been. known that when the colours with which fabrics are dyed are not the so-called fast colours, they fade in the light, that is, become paler. In this case the colouring matter changes into a colourless or differently coloured body; and that this is the effect of light is evident from the fact that those parts of the material in question which are covered up from the light —beneath the folds, for example—remain unchanged. This effect of light upon colour has been long turned to practical use in the bleaching of linen. The unbleached fabric is spread out in the sunlight, and repeatedly moistened with water; and thus, through the combined effect of light and moisture, the dark colouring substance becomes gradually soluble, and can then be removed from the linen by boiling it in alkaline lye.

It was formerly believed that the changes we have just described were caused by the heat which is produced in bodies by the sun's rays. That this is an erroneous

view is evident from the fact that fabrics dyed in colours
which are not fast can be exposed for months together
in the temperature of a hot oven without any bleaching
effect; and further, that wax, which the sunlight likewise bleaches, becomes darker, rather than paler, through
heat.

As we remarked before, the bleaching effect of sunlight is a slow process, and this circumstance renders the
phenomenon less striking.   A sudden and rapid occur-
rence surprises us, and stirs us up to inquire and to
reflect.

In the mines of Freiberg is occasionally found a
vitreous dull-shining silver ore, which, on account of its
appearance, is called horn silver (chloride of silver).

This horn silver consists of silver and chlorine in
chemical combination, and can be artificially produced by
passing chlorine gas over metallic silver. This horn silver
in its original position is completely colourless, but as
soon as it is exposed to the daylight it assumes, in a few
minutes, a violet tint. This effect of light has long
excited the astonishment of men of science.

In another substance containing silver this phenomenon is still more apparent.   Silver, placed in nitric
acid, is dissolved with effervescence, and if the solution is
evaporated, a solid mass of crystals is obtained. This is
not silver, but a combination of this body with nitric acid.
This nitrate of silver is totally different from ordinary
silver; it is easily soluble in water; it has a bitter, disagreeable taste; it fuses readily and destroys organic
matter; and it is therefore used as a corrosive agent,
under the name of lunar caustic.

It has been long known that the fingers which have

grasped lunar caustic, skin which has been cauterized by it, or anything sprinkled with a solution of it, quickly assume a dark colour. This can be at once observed by moistening a small piece of paper with a silver solution, allowing it to dry, and then placing it in the light.

These properties were soon made use of to produce a so-called indelible ink, which is nothing more than a solution of one part of nitrate of silver in four parts of water, mixed with a thick solution of gum. Written characters traced with it upon linen cloth are pale; but, when dried in the sunlight, quickly become dark brown, and are not injured by washing. Ink of this kind is much used in hospitals for marking linen. A quill, not a steel pen, must be used, as steel decomposes the nitrate of silver. It is not unusual to print the characters by means of wooden type.

From the discovery of the blackening of paper saturated with lunar caustic to the invention of photography there was but a step; yet it was long before any one thought of producing pictures by the help of light alone, and still longer before these attempts were crowned with success.

Wedgwood, the son of the celebrated manufacturer of porcelain who produced the popular Wedgwood ware, and Davy, the celebrated chemist, made the first attempts in the year 1802. They placed flat bodies, such as the leaves of plants, upon paper prepared with nitrate of silver. Light was thus kept from the parts of the paper covered by the objects, these parts remained white, whilst the uncovered portions of the paper were blackened by the light; and thus was produced a white outline, or "white silhouette," of the superimposed objects upon a black ground.   (See Figs. 1 and 2.)

Fig. 1.   Fig. 2.

Wedgwood produced in this manner copies of drawings upon glass, in white lines upon a black ground; and this process became the basis, in modern times, of a mode of treatment which attained the highest importance, coming under the name of the *lichtpaus* process.

Unfortunately these pictures were not durable.   They had to be kept in the dark, and could only be exhibited in a subdued light.   If they remained long exposed to the light, the white parts also became black; and thus the picture disappeared.   No means were then known to make the pictures durable, that is to say, to make them unalterable by light, or, as we now say, to *fix* them.   But the first step towards the discovery of photography was made; and the idea of producing pictures of objects without the help of the draughtsman became, after these first attempts, so extremely attractive that, from that time, both in England and in France, a large number of persons occupied themselves with the subject in private with the greatest enthusiasm.

It is clear that by the process of Wedgwood and Davy only flat bodies could be copied, and, notwithstanding all the improvements of which the process was still susceptible, it admitted of only a limited application.

But Wedgwood had already conceived the idea of the possibility of producing pictures of any bodies whatsoever by the action of light on sensitized paper. He tried to effect this by the aid of an interesting optical instrument which has the property of forming flat images of solid objects. This instrument is the camera obscura.

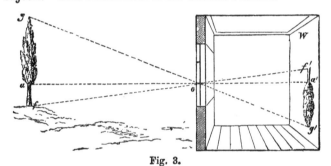

Fig. 3.

If a small hole be made in the window shutter of a completely darkened room on a sunny day, a clear image of the landscape will be seen on the opposite wall of the room.

Let a be a poplar, o the hole, and w the back wall of the room, then from each point of the poplar rays of light will travel towards the hole, and beyond that in a straight line to the wall. It is now clear that to the point a' light can only arrive from the point a of the poplar, which is situated on the extension of the line $a'o$. Therefore this point of the wall can only reflect light, which in its colour and position corresponds to the point a. The same remark applies to the points f and g, and the result accordingly is that on the wall an inverted image of the tree is visible. This was first observed by Porta, the celebrated Italian physicist of the 16th century, whose house, we are told by contemporaries, was seldom free from visitors in search of knowledge. An improved instrument was soon obtained by using instead of the room a small box (Fig. 4) which had a movable semi-transparent screen in place of the solid wall. On this screen

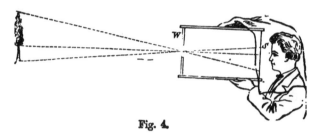

Fig. 4.

the image of an object in front of the box is clearly visible, if a minute hole is made in the front side, which is best if composed of a thin tin plate.*

These images appear still more beautiful if, instead of a hole, a glass lens, or, as it is called, a burning-glass, is substituted. This lens, or "burning-glass," at a certain distance, which is equal to that of its focus, casts a distinct image of objects—which is much better defined and clearer than that which is produced by a hole.

* To prevent the access of light the head must be covered with a cloth.

In this improved form the instrument was employed by Wedgwood and Davy. Their idea was to fix on sensitized paper the image produced upon the screen. They fastened a piece of paper saturated with a silver salt upon the place of the image, and left it there for several hours—unfortunately without result. The image was not bright enough to make a visible impression upon the sensitized paper, or the paper was not sufficiently sensitive. It now became necessary to find a more sensitive preparation to catch the indistinct image; and this was achieved by a Frenchman—Nicophore Nièpce. He had recourse to a very peculiar substance, the sensitiveness of which to light was before unknown to any one—asphalt, or the bitumen of Judæa. This black mineral pitch, which is found near the Dead Sea, the Caspian, and many other places, is soluble in ethereal oils—such as oil of turpentine, oil of lavender—as well as in petroleum, ether, etc. If a solution of this substance is poured upon a metal plate, and allowed to cover the surface, a thin fluid coating adheres to it, which soon dries and leaves behind a thin brown film of asphalt. This film of asphalt does not become darker in the light, but it loses by light its property of solubility in ethereal oils.

If such a plate, therefore, is put in the place of the image of the camera obscura, the asphalt coating will remain soluble on all the dark places (shadows) of the image, whilst the light spots will become insoluble. The eye, it is true, does not perceive these changes; the plate, after being exposed to the influence of light, appears the same as before the exposure. But if oil of lavender is poured over the coating of asphalt, it dissolves all the spots that had remained unchanged, and leaves behind all those that had been changed by light, that is, had been rendered insoluble. Thus, after several hours exposure in the camera obscura, and subsequent treatment with ethereal oils, Nièpce succeeded, in fact, in obtaining a picture. This picture was very imperfect it is true, but interesting as a first attempt to fix the images of the camera, and still more interesting as evidence that there are bodies which lose their solubility in the sunlight. This fact was again made use of long after the death of Niepce, and it led to one of the finest applications of photography, that of heliography, or the combination of photography with copper-plate printing, which Nièpce himself, to all appearance, had already known.

A copper-plate print is produced thus:—A smooth plate of copper is engraved with the burin (or graving tool); that is to say, the lines which should appear black in the picture are cut deeply in the plate. In producing impressions, ink is first rubbed into these cuts, and then a sheet of paper is placed upon the plate and subjected to the action of a roller press, whereby the ink is transferred to the paper and produces the copper-plate impression. Nièpce endeavoured to utilize light in producing these engraved copper plates in place of the laborious process of cutting. To effect this he covered the copper plate with asphalt, as before stated, and exposed it to the light beneath a drawing on paper. In this case the black lines of the drawing kept back the light; and accordingly, in these places the asphalt coating remained soluble; under the white paper, on the contrary, it became insoluble. Therefore, when lavender oil was afterwards poured over the plate, the parts of the asphalt which had become insoluble adhered to the plate, whilst the soluble parts were dissolved and removed; and the plate in those places was laid bare. Thus a film of asphalt was obtained on the plate in which the original drawing appeared as if engraved.

If a corrosive acid is now poured on such a plate, it can only act on the metal in those places where it is not protected by the asphalt; and in such places his metal plate was in fact eaten into. Thus an incised drawing upon a metal plate was produced by the corrosive action of the acid, and a plate was obtained which, after cleaning, could be used for printing like an engraved copper plate. Copper-plate prints of this kind have been found amongst the papers left by Nièpce, which he called "heliographs," and showed to his friends as far back as 1826. This method, in an improved form, is still in use at the present day, especially in the printing of paper money, when it is requisite to produce a number of engraved plates which are all to be absolutely alike, so that one piece of paper money may perfectly correspond to another, and may therefore be distinguished from counterfeits. In this way the arms and the inscription on the face of the Prussian ten-thaler notes are printed off from heliographic plates. Thousands of people carry photographic impressions in their pocket-books, without knowing it. Nor is there any occasion to fear that these notes can be imitated easily by the help of photography or heliography. We shall show later on, that the ground tint, the paper itself, and the colour of the inscription present well-devised obstacles to all such imitations, and make them very difficult, if not impossible.

Niepce's impressions were undoubtedly very imperfect, and therefore remained unnoticed. He himself gave them up, and again entered upon a series of experiments to fix the charming images of the camera obscura. In 1829 Daguerre joined him; and both carried on experiments in common until 1833, when Nièpce died without having obtained the reward of his long-continued efforts. Daguerre went on with the experiments; and he would not, perhaps, have got further than Nièpce if a fortunate accident had not worked in his favour.

He made experiments with iodide of silver plates, which he produced by exposing silver plates to the vapour of iodine, a peculiar and very volatile chemical element. Under this treatment, the silver plate assumes a pale yellow colour, which is peculiar to the compound of iodine and silver. These plates of iodide of silver are sensitive to light, they take a brown colour when exposed to it, and an image is produced upon them when they are exposed to the action of light in the camera. A very long exposure to light, however, is necessary to this end; and the thought could scarcely have arisen of taking the likeness of any person in this manner, since he would have been obliged to remain motionless for hours.

One day Daguerre placed aside as useless, in a closet in which were some chemical substances, several plates that had been exposed too short a time to the light, and therefore as yet showed no image. After some time he happened to look at the plates, and was not a little astonished to see an image upon them. He immediately divined that this must have arisen through the operation on the plates of some chemical substance which was lying in the closet. He therefore proceeded to take out of the closet one chemical after the other, and placed there plates which had been exposed to the light, when, after remaining there some hours, images were again produced upon them. At length he had removed in succession all the chemical substances from the closet; and still images were produced upon the plates. He was now on the point of believing the closet to be bewitched, when he discovered on the floor a dish containing mercury, which he had hitherto overlooked. He conceived the notion that the vapour from this substance—for mercury gives off vapour even at an ordinary temperature—must have been the magic power which produced the image. To test the accuracy of this supposition, he again took a plate that had been exposed to light for a short time in the camera obscura, and on which no image was yet visible. He exposed this plate to the vapour of mercury, and, to his intense delight, an image appeared, and the world was enriched by a most beautiful discovery.

CHAPTER II.

THE DAGUERREOTYPE.

Its Publication and extended Use—Method—Improvements—Discovery
of the Portrait Lens—Æsthetic Effects of the Daguerreotype.

MANY persons at the present day, who have before their
eyes the grand productions of paper photography, such,
for example, as life-size portraits, view doubtless with
pity, or even contempt, the little pictures called daguerreotypes, from their inventor. The appearance of these
pictures is no doubt injured by the ugly mirror-like
surface which prevents a clear view of them. No such
objections were felt in the year 1839, when Daguerre's
discovery was first spread abroad by report. Pictures
were said to be produced without a draughtsman by the
operation of the sun's rays alone. That was of itself
wonderful; but it was still more wonderful that, by the
mysterious operation of light, every substance impressed
its own image on the plate. How many extravagant
hopes, how many evil prognostications, were associated
with the report of this mysterious invention ?

It was prophesied that painting would come to an
end, and that artists would die of starvation. Every one

hoped that he could himself prepare images of any objects which he desired.

A friend is leaving home: in an instant his image is permanently retained at the moment of departure. A joyous company is assembled: a picture is taken of it at once as a souvenir. All objects were to be thus retained as pictures by the chemical action of light: the landscape glowing with the magical effects of sunset, the favourite spot in the garden, the daily life of the streets—men, animals, everything.

Then came sceptics who declared the whole thing impossible. These persons were reduced to silence by the testimony of Humboldt, Biot, Arago, the three celebrated scientific men to whom Daguerre disclosed his secret in 1838. The excitement grew. Through the influence of Arago an application was made to secure to Daguerre a yearly pension of 6000 francs, provided he made public his discovery. The French Chamber of Deputies agreed; and, after a long and tiresome delay, the discovery was at length disclosed to the expectant world.

It was at a memorable public *séance* of the French Academy of Sciences in the Palais Mazarin, on the 19th of August, 1839, that Daguerre, in the presence of all the great authorities in art, science, and diplomacy, who were then in Paris, illustrated his process by experiment.

Arago declared that "France had adopted this discovery, and was proud to hand it as a present to the whole world;" and henceforth, unhindered by the quackery of mystery, and unfettered by the right of patent,*

* It was only in England that the process was patented, before its publication on the 15th of July, 1839.

the discovery of Daguerre made the round of the civilized world.

Daguerre quickly gathered round him a number of pupils from all quarters of the globe; and they transplanted the process to their homes, and became in their turn centres of activity, which daily added to the number of disciples of the art.

Sachse, a dealer in art still living in Berlin, was initiated into Daguerre's discovery on the 22nd of April, 1839, and was appointed Daguerre's agent in Germany. On the 22nd of September, four weeks after the publication of the discovery, Sachse had already produced the first picture at Berlin. These pictures were gazed at as wonders, and each copy was paid for at the rate of from £1 to £2; while original impressions of Daguerre fetched as much as 120 francs. On the 30th of September Sachse made experiments in the Park of Charlottenburg, in the presence of King Friedrich Wilhelm the Fourth. In October the earliest Daguerre apparatuses were sold in Berlin. The first set of apparatus was purchased by Beuth for the Royal Academy of Industry at Berlin, and is still to be seen there. After the introduction of the apparatus, it was in the power of every one to carry out the system; and a great number of daguerreotypists started into existence. Men of science, too, cultivated (more than they do now) the new art: among others, the physicists, Professors Karsten, Moser. Nörrenberg, Von Ettinghausen—nay, even ladies, as Frau Professor Mitscherlich.

The first objects photographed by Sachse were architectural views, statuary, and paintings, which for two years found a ready sale as curiosities. It was in 1840 that he first represented groups of living persons, and in this way photography became especially an art of portraiture. It made the taking of portraits its principal means of support, and in two years there were daguerreotypists in all the capitals of Europe.

In America a painter, Professor Morse, afterwards the inventor of the Morse telegraph, was the first to prepare daguerreotypes ; and his coadjutor was Professor Draper.

Let us now consider more closely the process employed in producing daguerreotype plates. A silver plate, or in the place of it a silver-plated copper plate, serves to receive the image. It is rubbed smooth by means of tripoli and olive oil; and then receives its highest polish with rouge and water and cotton-wool. It is only a perfectly polished plate that can be used for the process. This burnished plate is placed with its polished side downward upon an open square box, the bottom of which is strewn down with a thin layer of iodine. This iodine evaporates, its vapours come into contact with the silver, and instantly combine with it. By this means the plate first assumes a yellow straw-colour, next red, then violet, and lastly blue. The plate, protected from the light, is then placed in the camera obscura, where the image on the ground-glass slide is visible, and " exposed " for a certain time. It is afterwards brought back into the dark, and put into a second box, upon the metal floor of which there is mercury. This mercury is slightly warmed by means of a spirit lamp. At first no trace of the image is visible on the plate. This does not appear until the mercury vapour is condensed on those parts of the plate which have been affected by the light;

C

and this condensation is in proportion to the change which the light has caused. During this process the mercury is condensed into very minute white globules, which can be very well discerned under the microscope. This operation is called the development of the picture.

After the development the remaining iodide of silver, which is still sensitive to light, must be removed to render the image durable, that is, "to fix it." This is effected by using a solution of hypo-sulphite of soda, which dissolves the iodide of silver. Nothing more is required after this than to wash with water and dry, and the daguerreotype is completed. Sometimes, in order to protect the picture, it was usual to gild it. A solution of chloride of gold was poured over, and then it was warmed; a thin film of gold was deposited, which contributed essentially to the durability of the pictures. A picture of this nature, however, is easily injured, and requires the protection of frame and glass.

Daguerre's first pictures needed an exposure of 20 minutes—too long for taking portraits. But soon after it was found that bromine, a rare substance, having many points of resemblance with iodine, employed in combination with the latter, produced much more sensitive plates, which required far less time, perhaps not more than from one to two minutes, for exposure.

Many of us, perhaps, still remember the early period of photography, when persons were obliged to sit in the full sunlight, and allow the dazzling rays to fall directly upon the face—a torture which is clearly marked and visible on the portraits still preserved of these photographic victims, in the blackened shadows, the distorted muscles, and the half-closed eyes. These caricatures could certainly not bear any comparison with a good drawing from life, nor probably would portrait-photography have ever succeeded if it had not become possible to obtain good results in a moderated light. This was attained by the invention of a new lens—the double objective portrait lens of Professor Petzval, of Vienna.

This new lens was distinguished by the fact that it produced a much brighter picture than the old lens of Daguerre, so that it was now possible to take less dazzlingly illuminated objects. This lens was invented by Petzval in 1841. Voigtländer ground the lens according to his directions, and soon one of Voigtländer's lenses became indispensable to every daguerreotypist. By employing bromide of iodine and Voigtländer's lens, the exposure was reduced to seconds.

The daguerreotype art had therewith reached its zenith. However delicate the pictures so produced appeared, it was found, after the first enthusiasm had gone and had given place to a cold spirit of criticism, that much still remained to be desired.

First, the glare of the pictures made it difficult to look at them. Then there were several marked deviations from nature: yellow objects often produced little or no effect, and appeared black; on the other hand, blue objects, which appear dark to the eye, frequently, though not always, came out white.

This is still the case in photography, only now the attempt is made to diminish this defect by subsequent treatment of the plate (by re-touching the negative).

But still a well-grounded æsthetic objection was brought against these pictures.

It was indisputable that the daguerreotype greatly

surpassed painting in the wonderful clearness of detail,
and the fabulous truthfulness with which it reproduced
the outlines of objects. The daguerreotype plate gives
more than the artist, but for that very reason it gives too
much. It reproduces the subordinate objects as faithfully
as the principal object in the picture.

Let us take the simplest case—a portrait. A painter,
when he paints a portrait, does not by any means paint
all that he sees in nature. The original wears, perhaps,
a shabby coat, which shows a good many creases, perhaps
a spot of grease, or a patch; but this does not disturb
the painter in the least, for he leaves out these accidental
details. In the same spirit, if the original is seated
before an old whitewashed wall, the artist by no means
put such a wall into his picture, for he can leave out all
that is displeasing, or add, on the contrary, what he
wishes.

It is different in photography. In taking portraits,
all those minor accessories which disturb the picture are
reproduced as faithfully as the principal object in it—
the individual himself. Another point must be added to
this. The different elements admitted by the painter
into his picture are by no means made equally prominent.
The head is the chief consideration in every portrait.
The painter accordingly gives his best skill and care
to the painting of the head in the most careful manner.
At the very least he puts the head in the strongest
light, and leaves the rest of the picture in a half-shade.
But in a photograph it is by no means the head which
is generally the most prominent—frequently it is a chair,
or part of a background; and this detracts considerably
from the effect of the picture. Finally, the expression of

the face is an important point in a picture; and this varies with the mood of the sitter. A photograph gives the expression which the original had at the moment the picture was taken. Now, the expression varies, and is affected by a slight annoyance, a vexatious circumstance, *ennui*, or even by the motionless attitude which has to be observed during the process; and hence the portrait often looks strange and unnatural.

It is quite otherwise with painting. The painter has longer sittings of the original than the photographer; he soon learns to distinguish the accidental frame of mind from the characteristic expression of the face, and thus he is in a position to produce a portrait much more closely corresponding with the character of the original than that of the photographer can ever be. This naturally applies only to paintings of masters of the first order. In the portraits of the dauber, none of these advantages are found; and this large class disappeared, like bats before the light, when the art of sun-painting suddenly rose upon the world. Many of these themselves adopted the new art and attained to greater results than they could have done as painters.

The artist of merit has no cause to fear photography On the contrary, it proves advantageous to him by the fabulous fidelity of its drawing—through it he learns to reproduce the outline of things correctly—nor can it be disputed that, since the invention of photography, a decidedly closer study of nature and a greater truthfulness are visible in the works of our ablest painters.

We shall see further on, how even photography appropriated the æsthetic principles according to which painters proceed in preparing their portraits, and how

thereby a certain artistic stamp was given to these productions, which raised them far above those of the early
period. But this result was only possible when the
technical part of photography had been brought to perfection, and a material better adapted to artistic work
than an unyielding silver plate had been introduced.

CHAPTER III.

PAPER PHOTOGRAPHY AND THE *LICHTPAUS*, OR NEW
TALBOT PROCESS.

Talbot's Paper Photographs — *Lichtpaus* Paper — Leaf-prints—*Lichtpaus* Process and its Application.

IN the same year that Daguerre published his process for
the production of images on silver plates, Fox Talbot gave
to the world a process for preparing drawings on paper
by the help of light. Talbot was an English gentleman
of fortune, who, like many Englishmen of leisure and
means, employed his time in scientific observations. He
plunged paper into a solution of common salt, dried it,
and then put it into a solution of silver. In this way he
obtained a paper which was much more sensitive to light
than that employed by Wedgwood. He employed this
paper in copying the leaves of plants. Talbot himself
says, "Nothing gives more beautiful copies of leaves,
flowers, etc., than this paper, especially under the summer
sun; the light works through the leaves, and copies even
the minutest veins."

This is no exaggeration. In the possession of the
author there are prints of this kind, made by Talbot

himself, which show excellently the structure of the leaves.

The pictures copied in this way in the sunlight are naturally not durable, because the paper still contains salts of silver, and is therefore sensitive to light. But Talbot offered the means of fixing the pictures—he plunged them in a hot solution of common salt; in this

Fig. 5.

way the greater part of the silver salts was removed, and the pictures did not blacken to any considerable extent in the light.

The celebrated Sir John Herschel carried out this fixing process even more successfully by plunging the pictures into a solution of hypo-sulphite of soda. This salt, which dissolves all the salts of silver, was at that time very expensive, costing six shillings per pound. The production of this salt soon kept pace with the increasing demands of photography, and now it is offered for sale by the ton, and at as low a rate as sixpence the pound.

By this means the production of a durable sun-picture on paper, which Wedgwood had in vain attempted, was rendered possible. This method gave, indeed, only pictures of flat objects which could be easily pressed on paper; for instance, leaves of plants, patterns of stuffs, etc. The process has lately been resumed, after it had almost been forgotten. Charming ornaments of leaves, various plants, and flowers have been produced; and these copies are proportionally more beautiful than the earlier ones, because a much finer and more even-surfaced paper than that of Mr. Talbot has recently passed into trade, under the name of *lichtpaus papier.** We give on the accompanying page a faithful imitation of one of these leaf-prints which have become popular in America.

Since the sale of sensitive paper has rendered the production of these leaf-prints very easy, we give here the mode of producing them for our fair readers, who will be able in this manner, like their sisters in America, to make ornamental pictures for the adornment of lamp shades, portfolios, and similar things.

The leaves—such as ferns and the like—are suitably chosen, pressed between blotting-paper and dried, then gummed on one side and grouped gracefully by the fair artist upon a glass plate, in a small frame (Fig. 6). As

---

* This paper is produced by Mr. Romain Talbot, 11, Karlstrasse, Berlin.

soon as the whole is dry, the print can be at once commenced.*

A small piece of sensitized paper is placed on the leaves, the two wooden lids, h h, are laid upon it, and fastened down by means of two little cross-bar pieces of

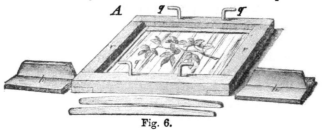

Fig. 6.

wood, x x, and then the whole is exposed to the light, the glass side uppermost. The sheet of paper very soon assumes a brown colour, where it is not covered by the leaves, and ultimately it receives a decided bronze tint. The light also penetrates partially through the leaves,

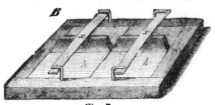

Fig. 7.

and colours the paper lying under them brown. It is easy to discern how far the paper under the leaves is

---

* These wooden frames, called printing frames, dishes, and fixing salt, are also manufactured by Mr. Talbot, at Berlin. There is now even a small plaything of this kind on sale, known by the name of the "sun-copying machine."

coloured if one of the cross-bars, x, and the half cover, h, are removed, and the paper is lifted up.

As soon as the impression is dark enough—it is quite a matter of taste whether the shade be dark or light—the paper is taken out and placed in a dark closet. Several pictures can, in like manner, be taken one after the other, and all of them can be afterwards fixed, that is, made permanent in the light, in one operation.

To this end the picture is placed in a flat dish (Fig. 8) containing water, for about five minutes, and then in a second dish in which is a solution of one part of hyposulphite of soda in five parts of water. The moment the print is dipped in this it becomes of a yellowish brown.

Fig. 8.

After remaining ten minutes in the fixing solution—several can be immersed in succession—it is taken out and placed in fresh water (most conveniently in a saucer). This operation of placing in fresh water is repeated from four to six times, the picture being left in the water three minutes each time.

Afterwards the pictures are placed on blotting-paper and suffered to dry; they can then be pasted upon cardboard, thick paper, linen, glass, or wood.

To many persons this process will appear only an agreeable pastime, but latterly it has gained an increasing importance as an aid in copying drawings, maps, plans, copper-plate impressions, and so forth.

This work of copying, which used to cost the artisan and artist many hours of time and labour, and yet was after all inaccurate, can be accomplished with the least possible trouble by the help of the process described above.

Let the reader imagine a drawing placed on a piece of sensitized paper and, after being firmly pressed by a glass plate, exposed to the light. The light penetrates through all the white spots of the drawing, and colours brown those parts of the paper lying under them; whilst the black lines of the drawing keep back the light, and thus the underlying paper remains white in these places. Therefore, if sufficient time is given for the operation of the light, a white copy on a dark brown ground is obtained in this manner, which is fixed and washed exactly like the leaf-prints described above. This copy is reversed with reference to the original, like an object and its reflection in a mirror; in other respects it is a faithful representation, stroke for stroke.

We give in Plate II. the copy of a woodcut struck off according to this method. This copy is but small, but the largest as well as the smallest drawing can be copied equally well; and copies of this kind, from drawings more than four feet square, are made in technical offices, in mines, and in machine manufactories.

Large printing frames are used for this purpose, resembling in their construction the small frames above described; and large wooden dishes, covered with a coating of asphalt, are employed for fixing and washing the prints. This operation is called in practice the *lichtpaus* process. The black copy is called a negative picture, but a second white copy can be prepared from this by placing the negative upon sensitized paper; then the light shines through all the white lines, and colours the paper lying under them of a dark hue, whilst it remains white under the dark places of the negative. In this manner a picture called a positive is produced, which perfectly resembles the original. The washing and fixing are carried out

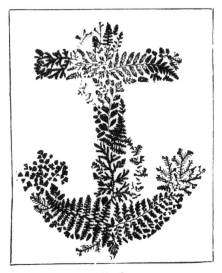

Fig. 9.

just as with the negative. Fig. 9 represents a positive of this kind taken from the negative, Fig. 5.

Thus the geographer is in a position to prepare quickly faithful copies of his sketches and maps, the engineer is able to copy the drawings of machines which are to serve for the workmen, and the student can copy illustrations of natural history which are to assist him in his studies.

In the process of copying, the sensitive paper must closely touch the original picture ; it must therefore be placed on the face of the picture.

This process has already done good service in military operations, where it was important to make quickly a copy of some map of which there was only one impression. If an attempt had been made to draw a copy of the map, it would have required several days to carry out, nor would the copy have been as correct as the print.

It is remarkable that this process, so important for industry, has only quite recently been known in its full value, although the experiments of Talbot have been before the world for more than thirty years. The explanation of this fact is, no doubt, to be found in the circumstance that the paper prints were then far less distinct than now, being often rendered worthless by spots and stains. Another reason is that the preparation of the paper requires especial care, and therefore frequently fails in the hands of the inexperienced; that is, of those who are not professional photographers. Further, the papers prepared according to the old method soon spoiled, and had on that account to be used immediately after their preparation.

These disadvantages have been removed by the invention of Romain Talbot's *lichtpaus* paper, which is sold ready prepared, and can be kept for months; and by this means the process can be easily nade available by every professional man and amateur.

CHAPTER IV.

THE DEVELOPMENT OF MODERN PHOTOGRAPHY.

Talbot's Paper Negatives—Photography as an Art for Multiplying Copies
—Services of Nièpce de St. Victor—White of Egg Negatives—Gun-
cotton in Photography—Collodion—Archer's Negative Process—
White of Egg Paper—Carte de Visite—Photographic Album.

THE reader has learnt in the previous chapter what a
negative is, and how by its means photographic copies of
flat objects can be obtained.

Talbot, the inventor of this paper process, carried out
further researches, in the endeavour to represent on paper,
by the help of the camera obscura, objects which cannot
be pressed upon sensitized paper; for example, a person
or a landscape.

He attained this object two years after Daguerre's
discovery, by means of paper prepared with iodide of
silver.

He dipped paper in a solution of nitrate of silver, and
then in a solution of iodide of potassium. He thus obtained a slightly sensitive paper, but one that could always
be rendered very sensitive, by plunging it into a solution
of gallate of silver.*

* The nature of this peculiar process is explained further on.

When this paper was exposed to the light in the camera obscura, a picture was not at once formed—this was only clearly developed after lying some time in the dark, or by subsequent treatment with gallate of silver —but it came out as a negative, and not as a positive. Thus, for example, in taking a portrait, the shirt appeared black, also the face; while the black coat, on the contrary, came out white.

The picture was fixed by plunging it in a solution of hypo-sulphite of soda.

The negative thus obtained is a picture on a plane surface of a solid object, and Talbot prepared positive pictures from negatives of this kind.

He placed the negative upon a piece of sensitive paper prepared with chloride of silver, as described in the last chapter, and allowed the light to act upon it. This shone through the white places of the negative, and imparted a dark colour to those parts of the sensitive paper lying under them, while the dark places of the negative protected the paper lying under them from the effects of the light. Thus he obtained a positive picture from a negative. He could repeat the process as often as he pleased, and therefore was in a position to prepare, by the aid of light, many positives from a single negative. Photography was thus classed among the arts that multiply copies, and this circumstance exercised an important result on its future development.

Daguerre's method only gave a single picture at a time; if more were required, the person had to sit several times. In Talbot's method a single sitting sufficed to produce hundreds of pictures.

It must be admitted that the earlier pictures of the

Talbot process were not remarkably engaging. Every roughness of the paper and each small speck of dirt were imprinted on the positive, which could not be compared in point of delicacy with the fine daguerreotypes ; but the method was soon improved.

Nièpce de St. Victor, nephew of Nicophore Nièpce the friend of Daguerre, conceived the happy idea of substituting glass for paper in the preparation of the negative. He coated glass plates with a solution of albumen in which iodide of potassium was dissolved.

A solution of this kind can be easily prepared by beating up white of egg to the consistency of snow, and allowing it to settle. The glass plates, after being coated with this solution, were dried, and then dipped into a solution of nitrate of silver. Iodide of silver was formed in this manner—the coating turned yellow, and became very sensitive to light.

Nièpce put these glass plates in the place of the image in the camera obscura, and suffered the light to act upon it.

No change was at first visible, but became clearly perceptible when the picture was immersed in a solution of gallic acid. Thus Nièpce obtained a negative on glass without the blemishes which appeared on the paper negatives.

He prepared prints from this negative by exactly the same process that had been employed by Fox Talbot, and he obtained from the fine negative a correspondingly fine positive, which was much better calculated to bear a comparison with the productions of Daguerre.

Nièpce invented his method in 1847. It excited much attention, but had its drawbacks : the preparation of the

D

albumen and the treatment with salts of silver and gallic acid was a dirty process. Therefore the method appeared to those who had been accustomed to the daguerreotype, dirty and unpleasant, and many were deterred from trying it.

On the other hand, the advantages of the new process in multiplying prints were so evident that it could not be overlooked; therefore, even those who had an objection to soiling their fingers nevertheless zealously devoted themselves to the work.

The readiness with which albumen decomposes was, however, a great disadvantage in the new process. They sought to avoid this by adopting a more durable substance.

This was afforded by the discovery of gun-cotton made by Schönbein and Böttcher in 1847. Schönbein found that ordinary cotton-wool dipped in a mixture of nitric and sulphuric acids assumes explosive properties similar to those of gunpowder. It was conceived that this substance would be an important substitute for gunpowder, but it was soon found that its explosive property was very unequal, being sometimes too strong and at other times too weak. On the other hand, another very useful property of this substance was observed—its solubility in a mixture of alcohol and ether. This solution leaves behind it a transparent membrane forming an excellent sticking-plaster for wounds. Thus the same substance that was destined to be a substitute for gunpowder, as a destructive agent for producing wounds, became actually a remedy for the latter. This solution of gun-cotton was called collodion.

The thought occurred to several photographic experimenters to try this substance instead of the white of egg, as a coating for glass plates; but the attempts did not at first lead to any satisfactory results. At length Archer published in England a full description of a collodion negative process surpassing in the beauty of its results, in simplicity and certainty, Nièpce's white of egg process.

Archer coated glass plates with collodion, in which an iodide had been dissolved; he immersed this in a solution of nitrate of silver, and thus obtained a membrane of collodion impregnated with sensitive iodide of silver, which he then exposed in the camera.

The invisible change produced by the light became visible on pouring gallic acid, or the still more powerful chemical agent, pyrogallic acid, over the plate; or, instead of this, a solution of proto-sulphate of iron.

Very delicate, clear negatives were obtained by this process, which yielded much more beautiful paper prints than the original paper negatives of Talbot. A very essential improvement was subsequently made in the preparation of photographic paper by coating it with white of egg, according to the process of Nièpce de St. Victor. This gave it a brilliant surface, and a warmer and more beautiful tone to the prints upon it, giving the pictures a brighter appearance than those produced upon the ordinary paper.

Thus Talbot's process, which at first seemed hardly worth notice compared with that of Daguerre, was gradually so perfected by successive improvements, that it ultimately took precedence of Daguerre's. After 1853, paper pictures from collodion negatives came more and more into vogue, whilst the demand for daguerreotypes fell off, and the production of the latter soon ceased altogether, except in some few places in America.

The collodion process is now the one universally employed. It acquired an immense impetus through the introduction of cartes de visite. These small portraits, which are intended to be given away, and therefore had to be produced in large numbers, were invented in 1858 by Disderi, the court photographer of the Emperor Napoleon, and obtained so great a success that they were immediately introduced into all circles, and soon became a necessity for everybody. The moderate price at which these portraits were sold made them attractive to the smallest purses, and the general public crowded to the ateliers, the number of which increased daily.

The old-fashioned album, the favourite souvenir of young people, was now superseded by the carte de visite, and the portraits of friends were substituted for their written words. A photographic album is now found in every home; and in Berlin alone there are at present more than ten photographic album manufactories, from whence they are exported to all parts of the world.

CHAPTER V.

THE NEGATIVE PROCESS.

The Dark Room—Chemically Inactive Light—Cleaning the Plate—
Application of Collodion—Sensitizing—The Camera—The Arrange-
ment of the Object—The Exposure to Light—The Development—
Intensifying—Fixing—Varnishing.

IN the previous chapters we have dwelt on the development of photography, and we are now able to feel at
home in the studio of a photographer. His whole business depends on the chemical action of light,—and yet
the scene of his principal activity is not the illuminated
studio, but a dungeon, in which the deepest night prevails, and which is called the dark room. The sensitized
plate, which has to be exposed to the light, and to
respond to its most delicate operations, must be prepared
in darkness, in the dark room. This space, surrounded
by bottles and pots, and crammed with instruments, is the
narrow world of the photographer, out of which he issues
only for a few minutes into the light of his studio, to
return directly with his exposed plate, and to subject this
to various chemical operations.

Many persons believe that the opening and shutting

of the cap—the cover of the lens of the apparatus, falsely
called the machine—is the chief work of the photographer. Nay, it is related of a certain queen, that she
thinks she is photographing, when she has all the necessary apparatus brought and prepared, and then, when all
is ready for the result, opens and shuts the cover of the
lens—a work that a child of five could do equally well.
But this operation is only one in a great chain of twenty-
eight operations, which each plate must undergo to
produce even a negative, while at least eight further
operations are required to prepare a positive from this
negative.

Let us look a little closer at these operations. The
appearance of a dark room is by no means inviting.
Even where the greatest order prevails, drops of silver
solution scattered about produce black spots here and
there. To this must be added a permanent odour of
ether from the evaporation of the collodion, and an unavoidable dampness from the necessary washing of the
plates,—and all this is seen in the hazy light of a gas or
petroleum lamp provided with a yellow shade, or of a
small window fitted with a glass of a similar colour.

The remark must here be made at the outset, that
the dark room of the photographer is not really completely dark. The daylight only must be excluded from
certain operations; but the yellow light of the lamp is
innocuous.

From this we learn the important distinction between
light chemically active, and light chemically inactive.
The light of the sun and of the blue heavens, the
electric and magnesium light, are chemically very
active, the light from gas or petroleum very slightly so;

whilst the yellow light of a spirit lamp, whose wick has
been rubbed with common salt, is entirely inactive. The
active light of day, furthermore, can be rendered inoperative if allowed to pass through a yellow or, better still,
a reddish-yellow glass. The light, therefore, that falls
through the yellow window of a dark room is chemically
inactive, or in so slight a degree active that it no longer
causes any disturbing effect. It is remarkable that yellow
light, which affects our eyes so powerfully, should have
no action on the photographic plate. Up to the present
time no satisfactory explanation of this fact has been
given. It has its disadvantages in practical photography;
for example, a yellow garment, a yellow complexion,
yellow spots—such as freckles—appear almost black in
the picture. Nevertheless, these disadvantages can be
obviated by retouching the negative as described at a
future page. On the other hand, this property of yellow
light has also its advantages for the photographer. It
permits him to prepare the sensitive plates in a light
which does not injure them, and yet enables him to
control the work. If the plates were sensitive to all
kinds of light, it would be necessary to prepare them in
absolute darkness, which would be very inconvenient.

The first operation required in preparing a sensitive
plate—an operation which requires great care—is the
cleaning of the glass. The plates, after being cut by the
diamond, are placed some hours in nitric acid, which
destroys all impurities adhering to the surface. The
acid is removed by washing, and the plate is then
dried with a clean cloth. To the uninitiated it would
then appear perfectly clean, but the photographer subjects
it to further polishing, by rubbing with a few drops

of spirits of wine; or, still better, of ammonia. The
slightest touch with the finger or rub of the sleeve, the
smallest drop of saliva which might chance to escape
from the mouth in coughing, would spoil the polished
surface; nay, even the atmospheric air produces with
time the same effect. If a cleaned plate is left only
twenty-four hours in the air, it gradually condenses
vapours on its surface, and another cleansing is rendered
necessary.

The cleaned glass is then coated with collodion. The
collodion itself is, as we know, a solution of gun-cotton
in a mixture of alcohol and ether, in which certain iodides
and bromides—for instance, iodide of potassium and
bromide of cadmium—have been dissolved. This solution must also be prepared with the greatest attention to
cleanliness; thus the purity of the materials employed is
of the greatest importance. The solution must be allowed
to stand a long time, and carefully poured off from any
sediment. The coating of a plate with collodion requires
a certain manual dexterity, and only succeeds with those
who have witnessed the process and after some practice.

It is usual to hold the plate horizontally by one corner
and to pour over the centre of it a pool of the thick fluid
and then to allow this to flow to all of the four corners
by a gentle inclination of the plate, ultimately pouring
off the superfluous fluid at one of the corners.

A considerable part of the fluid originally poured
upon it remains, and adheres to the plate.

Whilst pouring off the excess of collodion streaks are
liable to be formed, which would spoil the picture; to
avoid this the plate, whilst being drained, must be constantly kept in motion until the last drop has run off.

The fluid stiffens into a soft, moist, spongy film.   At this moment the plate must at once be immersed in the solution of nitrate of silver (called the silver bath).

And now a peculiar action of the fluids takes place, for the ether in the collodion film repels like fat the aqueous solution of silver, and a steady movement of the plate in the bath is necessary in order to make the solution adhere to the film.

This mechanical operation is accompanied simultaneously by a chemical change.   The iodides and bromides in the film undergo a double decomposition with the nitrate of silver of the bath, with formation of iodide and bromide of silver and nitrates of the metals previously combined with the iodine and bromine.   The iodide and bromide of silver colour the film yellow; and now the plate is ready to receive the picture to be painted by the light.

All these operations must precede the taking the photograph, and they are begun, in fact, at the moment when the person enters the studio, and with proper management the plate is prepared before the arrangement of the object is concluded.

This arrangement is a labour of itself; and it is of a genuinely artistic nature.   The points to which the photographer has to attend include a natural and yet graceful attitude of the original; the choice of the side which presents the most advantageous aspect; the picturesque arrangement of the dress; the removal of inappropriate objects which ought not to appear in the picture; the addition of those that are suitable, such as a table, a cabinet, or a background; lastly, an appropriate direction of the light.   Only a few minutes can be devoted

to these arrangements, for people dislike long delays or
experiments; and the plate itself only remains sensitive
a short time, for it is wet with the adhering solution of
silver, which soon dries up, and the plate is then useless.

When the exposure has been accomplished—during
which the person being photographed must remain perfectly motionless—the sensitive plate is brought back into
the dark chamber.

For the purpose of transporting the plate, which must
of course be guarded very carefully from the daylight,
the photographer employs a flat case or frame (Fig. 10),
called the dark slide. At the back of this slide is a
door D opening upon hinges, and in front there is a
sliding shutter H. In the corners are fixed four silver
wires d, d, d, d, upon which the plate rests with its prepared side downwards, being held steadily in its place by
a spring ff, fastened to the door D. The plate is carried
in the dark slide to the camera and substituted for the
movable ground-glass screen upon which the image of
the object has been previously focused. After the exposure the dark slide is removed from the camera and
taken back to the dark room.

And now follows one of the most important operations,
the development of the picture. Upon the plate there is
as yet no trace of a picture visible. The action of the
light produces a peculiar change of the iodide of silver
which forms the important constituent of the film; this
iodide acquires through the light the property of attracting silver, if this is precipitated on the film; this precipitate is produced by the following operation. If a
silver solution is mixed with a very dilute solution of
proto-sulphate of iron, there results by slow degrees a

precipitate of metallic silver—not, however, as a shining
mass, but as a grey powder. Now, a small quantity of the
nitrate of silver solution from the bath always adheres to
the film. If therefore a solution of proto-sulphate of iron
is poured upon it, a silver precipitate is formed, and the
picture suddenly makes its appearance, owing to the
silver adhering to those parts affected by the light.

The features of a portrait that are first visible are the
lightest—the shirt, then the face, and lastly the black

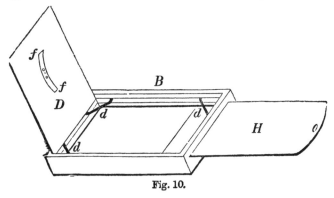

Fig. 10.

coat. The negative thus obtained, however, is by no
means completed by this operation.

The picture is usually too transparent to answer for
the production of paper prints; for the preparation of
such a print depends upon the light shining through the
transparent places of the negative, and colouring dark
the paper beneath, while it fails to penetrate the parts
which have to remain white. The opaque and transparent parts of the negative must be in sufficient contrast with each cther to produce this effect.

The negative must, therefore, be more strongly defined; and this is produced by repeating the developing process. A mixture of solutions of green vitriol and of nitrate of silver is poured upon the picture, a silver precipitate is again formed on it, which adheres only to the dark parts of the negative, giving them a greater density. If the plate is not perfectly clean, silver is precipitated, in the processes of developing and intensifying, upon the dirt stains and produces spots. After the further development of the picture, or the so-called intensifying process, has been completed, it is only necessary to remove the iodide of silver, which diminishes the transparency of the clear parts of the plate. For this purpose a solution of hypo-sulphite of soda is poured on the plate. This salt has the property of dissolving salts of silver, which are insoluble in water, so that the iodide of silver vanishes under the influence of this solution. This is the fixing process. Lastly, the plate is washed and dried.

If it be borne in mind that all these different operations are performed on a film liable to be injured by the slightest touch, it is not to be wondered at that the inexperienced beginner destroys so many coatings before he can prepare a perfect one.

Even when dried, the picture is very liable to injury; and therefore photographers, in order to protect it, cover it with a varnish, that is, with a solution of a resin, such as shellac or sandarach, in spirits of wine. The fragile glass negative is therewith brought to completion.

This sketch of the operations which a photographer is obliged to carry out in order to produce a negative, is sufficient to show that photography is a more difficult art than some persons imagine, and that it requires something more than the opening and shutting of a lid.

The chief requisite for the success of these operations is routine; accuracy is only attained in practising each part of the process. Faults that are made in any single operation are, as a general rule, irremediable; and therefore it is absolutely essential to avoid them.

CHAPTER VI.

THE POSITIVE PROCESS.

Character of the Negative—Departure from Nature—Retouching the Negative—Preparation of Sensitive Paper—Printing—Toning with Chloride of Gold—Fixing—Cause of Fading—Quantity of Silver in the Picture—Vignetting.

IN the preceding chapter we have become acquainted with the preparation of a negative. However interesting such a negative might be, nevertheless, it would not satisfy the purchaser of a portrait, since it shows everything reversed. The white face is black, and the black coat, white. No one would hang up on his wall a picture representing him as a Moor. It is therefore necessary to obtain a positive impression from this negative. We have already learnt how this is effected in the chapter on the *lichtpaus* process. It is the old Talbot method that is here employed. But we must still mention certain operations which have become very important in modern photography.

The camera, the negative process, and the photographer who knows how to manipulate intelligently, no doubt produce a negative which, laid over sensitive paper and exposed to the light, yields a positive; but although this positive is very faithful in the delineation of figures —that is, of their outline—it yet presents marked departures from nature. Thus the relations of light and shade are by no means correctly given. In general the light parts appear too light, the dark parts too dark—as, for example, the folds in a dress, the skin, and, moreover, the shadows under the eyes and chin. When photographers knew nothing of art, these defects were taken as a matter of course. People protested that photography was correct because nature, through photography, was herself the artist. But in this conclusion the co-operation of the photographer was overlooked.

No doubt nature, that is, the object to be taken, makes the impression upon the plate, by the light radiating from it; but an impression of light is not a picture—it is indeed, of itself, invisible; nay, more, the strength of the impression of light is entirely at the discretion of the photographer, who can make it weak or intense by a shorter or longer exposure. There is no rule which determines the length of time a photograph has to be exposed to the light.

The fact is that nature, probably speaking, only determines the outline of the picture, while the relations of light and shade depend partly on nature, and partly on the will of the photographer.

The impression made by the light must be developed to be visible, and finally the developed picture must be intensified. By this means the photographer can at his option increase, and even exaggerate, the contrasts of light and shade.

If the negative is carefully compared with the original, we shall find that many dark parts have not appeared at

all, because the exposure was too short for them to produce
an impression upon the plate ; others have appeared, but
too faintly.   On the other hand, very light parts—for
instance, the shirt-collar—are too apparent, and the
needlework upon it is invisible because the time of exposure was too long.   In the case of long exposures it is
often remarked that bright parts differing little in colour
are entirely confused and form a single white patch.

Moreover, the accessories which a painter would undoubtedly omit, such as warts, pockmarks, little hairs,
are all as clearly defined as the principal features; and
thus the negative is neither a correct nor an agreeable
repetition of the reality, but produces in the positive a
picture which shows considerable departure from nature,
and is often inaccurate by giving too much prominence to
accessories.

In the first period of photography these departures
were overlooked.   Every one was content to possess a
portrait which at least showed the outlines correctly;
and what was defective in the negative it was sought to
remedy by retouching the positive.   But this made the
pictures dear; and as it began to be the custom to order
pictures by the dozen, the endeavour was made to evade
this labour, which had to be applied to each individual
picture, by carrying it out in the negative.

A single retouched negative gave any number of
corrected prints which did not require to be retouched,
and thus retouching negatives became the first and most
important operation in producing a faithful and agreeable
picture.   The essential characteristic of this operation
consists in entirely covering many parts.   For example,
the freckles and warts which are white in the negative

are entirely removed by a pencil, or by spots of Indian ink. Other parts—for example, the ill-defined details of the hair—are brought out by pencil strokes. Many hard shadows, such as the wrinkles in the face, are softened off by slight touches of Indian ink. It must be constantly borne in mind during this operation that all the dark lines which the painter draws on the negative will appear light in the positive.

It is requisite, therefore, for successful retouching, to know how to use the pencil and brush so as to produce the desired effect in the positive. The best draughtsman or painter is therefore far from being qualified for retouching a negative.

It is to be remarked that the negative retouch may, under certain circumstances, go too far. By covering every wrinkle an old face can be made young; an ugly original can be beautified by cutting away a hump on the back, or other abnormal growths; and these tricks are often put into requisition for the vanity of sitters, and are dearly paid for.

Plate VI. represents two portraits of the same person, one from a retouched negative, the other from a negative that had not been retouched. The spots on the skin and the dark shadows, on the picture which is not retouched, are clearly to be seen, while in the retouched one they are not visible.

In many cases retouching is employed simply to satisfy human vanity, but this is by no means always the case.

As already explained, photography does not always represent the natural colours correctly. Yellow often becomes black, and blue, white. For this reason photo-

E

graphs of paintings frequently fail to give the various tones of the original. This fault may be in some degree corrected by retouching the negative, and it is owing to this alone that photographs of oil paintings have attained their present perfection. We will treat of this subject in a future chapter. Let us consider the preparation of a positive picture.

The first operation is the preparation of the sensitive paper.

A piece of paper coated with white of egg and moistened with a solution of common salt is laid in a dish on a solution of nitrate of silver, and allowed to remain for about a minute. The solution is absorbed by the floating paper, and chloride of silver is formed by double decomposition of the nitrate of silver with the common salt.

Fig. 11.

The wet paper is but slightly sensitive; it becomes fully sensitive only after being dried. The dry paper, saturated with chloride and nitrate of silver, is then pressed upon the negative in the printing frame (Fig. 11), which is similar to the one described. Then the whole is exposed to the light. The same process ensues which we have described in the chapter on *lichtpaus* paper; the light shines through the clear places of the negative and colours the paper lying under them dark, the paper under the dark places of the negative remains white, while it assumes a slight colour under the half-tones. In this manner a faithful positive copy of the

negative is produced, presenting a beautiful violet-brown
tint. We know from the description of the *lichtpaus*
process, that this print would not remain unaltered in
the light for long, because the paper is still sensitive to
light. The salts of silver contained in it must be removed
if the impression is to be made lasting. To this end a solution of hypo-sulphite of soda is employed. If the prints
be immersed in this solution, they become durable in the
light; but, unfortunately, they suffer a peculiar change
of colour, assuming an ugly brown tint. This tint is of
no consequence in technical and scientific pictures, but
detracts greatly from portraits and landscapes; and in
order to give these a more agreeable tint, before fixing
them, they are immersed in a dilute solution of chloride
of gold. This process is called toning.

In this operation gold is precipitated on the picture,
giving it a bluish shade; and now the tone of the picture
is not essentially altered by hypo-sulphite of soda.

The picture thus produced consists partly of gold,
partly of silver, in a finely divided state, and only requires
to be thoroughly washed in order to become perfectly
permanent. If this washing is omitted, small particles
of hypo-sulphite remain behind, which decompose and
form on the picture yellow sulphide of silver. This
accounts for the circumstance that the pictures of an
earlier period, when from ignorance of this fact this
thorough washing was neglected, so often became faded
and yellow.

It is surprising what a small amount of silver and
gold is required to give an intense colour to a whole
sheet of paper. For in a sheet of perfectly blackened
paper 44×47 centimetres there is only 0.15 gramme

of silver, and in a photograph of the same size only 0·075 gramme; in a carte de visite but 0·002 gramme.

It must be here remarked that prints bleach a little in the fixing process; and hence the photographer usually lets the prints become darker than they are to remain. Thus even the printing process requires a practised eye, simple as it may appear.

In certain cases tricks of art are employed to produce agreeable effects, and among these is that of vignetting. Our readers are no doubt well acquainted with portraits on a white ground, the outlines of which gradually become confounded with the ground tint of the picture. This effect is produced in a very simple manner by placing what is called a mask over the copying frame. This mask is a piece of metal or cardboard (Fig. 12) in which an oval hole b is cut. This is placed on the printing frame $K\,K$, so that the part of the negative which is to be printed lies perpendicularly under it. This part is

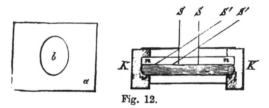

Fig. 12.

then acted upon by the broad perpendicular bundles of light $S\,S$, and intensely coloured, while the marginal parts lying under the mask are affected only by the narrow slanting pencils $S'\,S'$, and therefore are coloured less intensely in proportion to their distance from the hole of the mask. Thus a gently vanishing margin is produced, looking very artistic, and yet only the result of a very simple trick of art.

The picture produced in the manner described above only requires some preparation to be an elegant drawing-room ornament. It is cut in a regular shape, square or oval, and fastened with clean paste to white cardboard; after drying, any slight blemishes are removed with the paint brush, and finally its surface is rendered glossy by two polished steel rollers.

Certain sizes have been adopted by the public, such as the "carte de visite" and "cabinet" sizes. The former is rather larger than an ordinary visiting card. The latter is two and a-half times as large.

The carte de visite, introduced at Paris by Disderi in 1858, speedily secured admirers, and has spread over the whole earth. Even Chinese photographers have adopted the carte de visite form.

The carte de visite and the cabinet size—which was first introduced in England, and is a great favourite in America—are not confined to portraits, but also employed for landscapes and photographs taken from oil paintings. Millions of these pictures are sold every year, and a properly arranged album for preserving them is found in almost every family.

Photography admits of such small pictures because of its fine details, but it is by no means confined to them. Surfaces on which life-size portraits may be taken can be employed. The production of the latter necessitates a peculiar process, called the enlarging process, which will be treated of at a future page.

CHAPTER VII.

LIGHT AS A CHEMICAL AGENT.

Theory of Photography—Nature of Light—Undulatory Theory—Action
of Light on Realgar—Chemical Decomposition by Light—Colours
and Tones—Their Vibrations—Refraction—Dispersion—The Spectrum—Lines of the Spectrum—Invisible Rays—Photographs of
Moonlit Landscapes—Abnormal Photographic Effect of Colours—
Photography of the Invisible.

" GREY is all theory, and green life's golden tree," says
Goethe. This saying has often been misunderstood and
misused, especially by those too lazy to think; but,
faithful to its true meaning, we have first treated of a
multitude of facts from life—that is, from the history
and practice of photography,—and now we proceed to
describe, by the help of science, how and why, not the
golden but the silver tree of photography grows, blooms,
and bears such splendid fruit.

Two sciences join hand to accomplish the wonders of
photography. One is Optics, a division of Physics, and
the other Chemistry. We have already shown that they
alone are inadequate to fulfil the requirements of photography. Æsthetical claims have to be considered; and
thus photography unites in itself the provinces of natural

science and of the fine arts which seem remote and incapable of union. We shall consider first the principles of optics—that is, light—as the force which occasions the chemical changes in photography. We shall see that its chemical operations have not only become the basis of our art, but that they have played, and still play, a much more important part in the development of our planet.

We are aware of the existence of sun, moon, and planets. We know their distance; nay more, we know their elements, though we are separated from them by millions of miles.

We are indebted for all this knowledge to light. What is light? A wave motion of ether. And what is ether? An infinitely delicate fluid, which fills all the space of the universe, and which, like all fluids, may be set in undulation. If we throw a stone into water, waves are produced—that is, circles or rings of hills and valleys, are formed, which appear to widen out from a centre, and as they extend become gradually less, until they finally disappear. If several stones are thrown at the same time into the water, each of them forms its own system of waves.

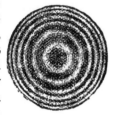

Fig. 13.

These intersect each other in the most intricate manner; and, although an apparent confusion of rings takes place, it is wonderful that none of them disturbs the other, and that each circle widens out regularly from its own centre, where the stone fell into the water.

If a handful of sand, which contains many thousand grains, is thrown into water, and if the attention be directed to the undulations of a single grain, it will be remarked that this one, without being affected by the countless other waves, widens out in a regular circle.

These undulations are one of the most remarkable movements in nature, taking place not only in water, but in the air, where they occasion the propagation of sound.

The peculiar feature of the undulatory movement consists in the fluid appearing to advance without really doing so. If, sitting on the side of a sheet of water, we see an undulation approach, it appears exactly as if the particles of water were approaching us from the origin of the movement.

It is easy to prove that this is an error by throwing sawdust or a piece of wood into the water. It dances up and down upon the ripples without moving from the spot. Indeed, the undulation is itself only an up and down motion of the particles of the water, and this movement is communicated further and further to the neighbouring particles.

Exactly in the same manner light spreads in undulations from a luminous body through the ether of space in all directions. The direction of the undulation we call a ray of light. We perceive it as soon as it reaches our eye, because the vibrating ether strikes our retina.

Now, we know that the undulations of sound are able to set other bodies in motion. If the A string of a violin is struck, the A string of a piano standing near sounds distinctly with it. Nay, even if the damper of a piano is raised and any note be sung, instantly the string of the piano sounds which has the same tone. The same thing happens with a glass bell of the same tone. There are people even who can break a glass by a shrill tone of their voice. The glass is so shaken by the violent undulations communicated to it by the air, that it falls to pieces. Under such circumstances, it need not surprise us that the undulations of light agitate bodies so forcibly that they fall to pieces.

Realgar offers the most remarkable example of this kind. This is a beautiful mineral of a ruby red colour, in the form of splendid crystals composed of sulphur and arsenic. If a crystal of this kind be exposed for months to the light, it falls into powder; and in this way many very fine pieces of this beautiful mineral have been lost in the mineralogical museum of Berlin.

This is only a mechanical, and not a chemical, operation of light; but it gives an insight into its chemical action. Heat occasions chemical decomposition by expanding bodies, and thereby removing their atoms so far apart that the chemical power which unites them loses effect, and the component parts separate. Thus oxide of mercury is by heat resolved into its constituent parts, mercury and oxygen.

Decomposition is effected by light when the atoms of a body are agitated by its undulations, that is to say, are made to vibrate; and if these vibrations are unequal, a separation of the parts takes place, and the body falls to pieces.

Light waves are not a fiction. Not only has their existence been ascertained, but their size has been determined. The latter is extremely minute, but nevertheless is susceptible of measurement.

The waves of sound and the waves of light have therefore a certain analogy; and as there are different notes in music, so are there different colours in light. The number of notes is great. The simplest piano now has nine octaves, and there are other tones below and above these. But the number of colours is small; only seven of them can be distinguished—red, orange, yellow, green, blue, dark blue, and violet,—the well-known colours of the rainbow. The painter, indeed, contents himself with three ground tints—yellow, blue, and red. All the others are the result of their mixture; and the large scale of colour of the painter consists not of simple tones of colour, but of what may be called chords of colour.

The deep tones of music are caused by few undulations, the higher tones by more. For example, an $\bar{a}$ string makes 420 vibrations in a second, the a an octave lower makes 210, the great A 105.

In light, red is the colour which gives the fewest vibrations; it is the lowest tone in colours, and violet is the highest, giving vibrations nearly twice as rapid as red. With regard to tones, we know that they all spread with equal rapidity in the air; if this were not the case, a piece of music would be heard in the distance as a most disagreeable discord.

It is the same in the kingdom of light—the colours, without exception, are propagated through the ether with equal rapidity, the red as fast as the violet. But, whilst sound passes over only 1100 feet in the second, light traverses 190,000 miles in the same time, and the deepest colour-tone—red—makes in a second 420 billion of vibrations; that is to say, a million times a million as many as the tone which is marked in music with a bar over the a,* that is,

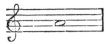

The small number of the colour-tones compared with the large number of musical tones is very striking. But the fact is, that, besides the seven visible colours, there exist invisible tones, which are both deeper and higher than the visible colours.

These invisible colour-tones are partly disclosed by the thermometer, which reveals the lower tones, and partly by substances sensitive to light. For it is remarkable that the colour-tones, which are higher than the violet, though invisible, have a powerful chemical effect.

We name the invisible tones of colour above violet, ultra-violet, and those below red, ultra-red.

In the common white light all the tones of colour are found together, and in combination they produce the effect of whiteness; but if we wish to examine the tones of colour separately, we must part them, and this may be done by the help of a prism.

Any polished crown-glass prism causes flames seen through it to appear like a rainbow containing the primitive colours we have named above. This separation of the colours in the prism takes place by refraction.

If a ray of light passes from one transparent medium

* We may here remark that the tone a is not everywhere the same. The a of the Berlin Opera is the highest; it has 437 vibrations,—the Italian Opera at Paris only, on the contrary, 424 vibrations. We have adopted for the sake of simplicity a round number, 420.

to another, it is deflected from its rectilinear direction, and this deflection is named refraction.

For example, if the ray $a\ n$ (Fig. 14) strikes a surface of water, it does not continue in its original direction $a\ n$, but in the direction $n\ b$. If through the point n, where the ray strikes the water, a perpendicular line $d\ f$ be drawn, the rule is that when a ray passes from a thinner medium (for example, air) into a denser one, it approaches the perpendicular, for $n\ b$ is evidently nearer to the $d\ f$ than $n\ a$. On the other hand, when a ray passes from a denser to a thinner medium,—for instance, from glass into air,—then the ray $n\ b$ departs from the perpendicular $n\ d$; that is, the angle which the ray makes with the perpendicular after refraction is greater than the angle which it makes with it before.

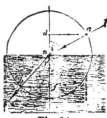

Fig. 14.

Now, it is a remarkable fact that light of different colours is refracted unequally.

If a bundle of rays of white sunlight is suffered to fall on a piece of glass, the violet rays are deflected more than the blue rays, the blue more than the green, yellow, and red ; and the result of this is that the white bundle is decomposed into a rainbow-coloured fan, violet, indigo, blue, green, yellow, orange, and red.

This phenomenon is the cause of the rainbow. If a ray a falls on a drop of water (Fig. 15), it is refracted and at the same time divided into a coloured fan, which is reflected from the back of the drop, suffers again refraction and dispersion at b, and issues as a broad bundle of colour. In open daylight this cannot be clearly seen, because our eyes are dazzled by the bright light. In order to observe the pure colours of the spectrum, it is best to produce it in a darkened room, in which the light is allowed to enter only through a small slit (*b* Fig. 16).

When a prism *S* is placed behind the slit, a pure spectrum appears upon the opposite wall. If the slit is sufficiently narrow, a series of dark lines may be observed within the spectrum, at right angles to it.

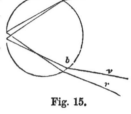

Fig. 15.

These lines were first seen by Wollaston, and studied more exactly by the celebrated optician Frauenhofer, and called after him Frauenhofer's lines.

The lines are always found in the same position, so that they can be considered as natural music lines, upon which the scale of colour is written ; and as the music lines serve for the recognition of the musical notes, so do the

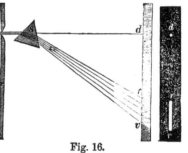

Fig. 16.

lines of the spectrum indicate fixed points in the scale of colour.

If we were to speak of the green of the spectrum, this would be a very vague designation; whereas by mentioning a line of the spectrum in the green, the part of spectrum is at once characterized. For this purpose Frauenhofer

gave the most characteristic lines names, the letters of
the alphabet; a certain line in the red he called *A*,
another in the yellow *D*, one in the violet *H*, and *H'*. As
the number of lines reaches several thousand, these
letters do not suffice to indicate them all. (See Fig. 17.)

The lines thus named are found in sunlight; the
light of other stars commonly shows other lines. The
light from artificial sources does not show dark, but
bright lines; a flame coloured yellow with common salt
shows, for example, a very characteristic line in the
yellow; a burning magnesium wire shows several blue
and green lines.

The situation of these lines agrees exactly with that
of certain dark lines in the spectrum. For example, the
yellow line in a flame coloured with common salt exactly
coincides with the line *D* in the spectrum. The green
lines in a flame of magnesium coincide exactly with lines
E and *b* in the spectrum.

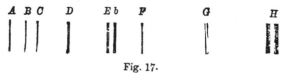

Fig. 17.

This remarkable coincidence led to the surmise that
the lines in the sun's solar spectrum might owe their
existence to the same substances that produced the
coinciding bright lines in flames. Kirchhoff converted
this surmise into a certainty, and was thus able to
determine from the lines in the solar spectrum the substances present in the sun, and thus to demonstrate the
chemical composition of a star distant more than 20
millions of miles (spectrum analysis).

But the spectrum contains still other wonders, which cannot be discerned by the human eye, but by the photographic plate.

If a sensitive plate be exposed to the action of the spectrum, it is observed that the red and yellow rays have scarcely any action, whilst that of the green is but very weak. Light blue produces more effect, dark indigo and violet the most; and in the space where no rays can be perceived by our eyes, a distinct action is produced, and extends beyond the violet for a space almost as long as the whole visible spectrum.

From this fact the existence of the ultra-violet rays was ascertained. The retina of our eye and the photographic plate possess an entirely different sensitiveness. Our eye is affected most powerfully by yellow and green light. These colours appear to us the brightest, while the photographic plate is not at all affected by them; on the other hand, indigo and violet rays, which appear dark to our eye, and even rays which to our eyes are invisible, produce a powerful action on the plate.

It is natural, therefore, that photography should represent many objects in a false light. Further back we called attention to the fact that photography is much less sensitive than the human eye to feebly lighted objects. This is most clearly seen in the fact that the eye can easily perceive objects by moonlight, which is 200,000 times weaker than that of the sun; whereas the photographic plate is not able to produce any picture of a moonlit landscape. The photographic landscapes by moonlight sometimes offered for sale have been taken in the daylight and copied very darkly, so that they produce the effect of moonlight. These pictures are very popular at Venice.

This small sensitiveness of the photographic plate to feeble light explains the reason why shadows are generally too dark in photographs. To these defects must be added the false action of light,—blue generally appears light, yellow and red, black. Yellow freckles appear therefore in a picture as black spots, and a blue coat becomes perfectly white. Blue (and therefore dark) flowers on a yellow ground produce, in photography, light flowers on a dark ground. Red and also fair golden hair becomes black. Even a very slight yellow shade has an unfavourable effect. A photograph from a drawing is often blemished by little iron-mould specks in the paper invisible to the eye. These specks frequently appear as black points. There are faces with little yellow specks that do not strike the eye, but which come out very dark in photography. A few years ago a lady was photographed at Berlin, whose face had never presented specks in a photograph. To the surprise of the photographer, on taking her portrait specks appeared that were invisible in the original. A day later the lady sickened of the small-pox, and the specks at first invisible to the eye, became then quite apparent. Photography in this case had detected before the human eye the pock-marks, which were, doubtlessly, slightly yellow.

In the photographs of paintings, such abnormal action of colour becomes still more evident, and can only be removed by appropriate retouching.

It is proper to observe, however, that by no means all shades of blue become light in photography. For example, indigo forms an exception, appearing as dark as in nature; this is shown in the photographs of the uniforms of Prussian soldiers. The reason of this is, that indigo contains a considerable amount of red. On the other hand, cobalt blue and ultramarine produce almost the effect of white. Again, vermilion becomes dark, also English red; whereas Turkey red, which contains blue, becomes very light. Chrome yellow becomes much darker than Naples yellow; Schweinfurt's green becomes lighter than cinnabar green. No one of our pigments is a perfectly pure spectrum colour, but consists always of a mixture of different colours, and therefore is essentially modified in photography.

If the effect of the colours of the spectrum on photographic plates is more narrowly examined, it is observed that the indigo produces the greatest action. Nevertheless, the differently sensitized photographic plates offer somewhat various results in this respect. Chloride of silver is most sensitive to violet, but non-sensitive to blue. Bromide of silver is sensitive even to green, and iodide of silver only to violet and indigo. Mixtures of iodide and bromide of silver are sensitive both to blue and green. The author succeeded, in the end of 1873, in making photographic plates sensitive even to those colours that were before considered to be inoperative, i.e., yellow, orange, and red. He found that if certain coloured substances that absorb green light were added to bromide of silver, which is by itself but slightly sensitive to green, the sensitiveness of this bromide to green is considerably increased. In like manner, the addition of coloured substances absorbing yellow or red light makes bromide of silver sensitive to yellow and red light. After this discovery, we may hope that the difficulties attending the taking of coloured objects may be soon overcome.

F

Mention has often been made of the photography of the invisible. The cases already recorded of the photographs of invisible pock-marks belong to this. But the photography of invisible quinine writing is especially understood by the term, photography of the invisible. If a writing is made on paper with a concentrated solution of di-sulphate of quinine, the result is scarcely visible. If this is photographed, it appears black and plainly visible in the picture. The di-sulphate of quinine has the property of lowering the tone of violet, of ultra-violet and blue rays; that is, of converting them into rays of less refractive power and of less chemical effect; therefore the light reflected from quinine produces little or no photographic effect, and the written characters become black.

This property of the di-sulphate of quinine serves also to make ultra-violet rays visible. If a piece of paper which has been moistened with a solution of sulphate of quinine is held in the spectrum, the originally invisible ultra-violet part is seen to shine with a bluish green light.

Other substances produce this effect, such as glass coloured with uranium and fluor-spar from Devonshire, and therefore this property has received the name of fluorescence.

CHAPTER VIII.

CHEMICAL EFFECT OF DIFFERENT SOURCES OF LIGHT.

Artificial Light—Magnesium Light—Lime Light—Electric Light—
Illumination of Subterranean Places by Reflected Sunlight—
Chemical Intensity of the Light of the Sun and of the Blue Sky
—Breathing of Plants under the Influence of Light—Effect of Light
in the History of the Development of the Earth and in the
Economy of Nature.

FROM the facts explained in the foregoing chapter, it
follows that chemical effects are chiefly produced by the
ultra-violet, violet, and blue rays. It is therefore evident
that the chemical action of any light will be proportional
to the amount of these rays it contains.

Lamplight (gas or petroleum) is very poor in such
rays, and therefore acts but feebly on the photographic
plate; photographers are thus enabled to prepare their
sensitive plates in a subdued lamplight.

This is also frequently done in the day by allowing
the light to pass through yellow glass.

The white Bengal light, the flames of the blue Bengal
light, and those of burning sulphur, produce a much more
powerful chemical effect. The latter possesses only a
small illuminating power, because it contains few yellow

and red rays; but, on the other hand, it is rich in blue
and violet. Photographs have been actually taken by
help of these flames.

But the chemical action of the foregoing lights is
greatly surpassed by the effect of the lime, the magnesium, and electric lights. The magnesium light is
very simply produced by the burning of magnesium wire.

Fig. 18.

Magnesium is a metal which forms the chief component part of magnesia. Magnesia is nothing but
magnesium rust; that is, a combination of magnesium
with oxygen.

If magnesium wire is burned, it combines with the
oxygen of the air, producing a brilliant light and
forming the oxide of magnesium. This light is very
convenient in its application. An ounce of magnesium

wire, sufficient to take fifteen to thirty photographs, can be easily carried in the pocket. The general use of the light is impeded by the price of the metal (sixpence per gramme) and by the smoke which it emits. The author has repeatedly employed it with success in taking the sculptures in the sepulchral monuments of Egypt. When burning the magnesium wire, Solomon's lamp is used (Fig. 18). This consists of a reel K, upon which the wire is coiled, a clockwork apparatus G, which conducts the wire from the reel between two rollers to the tube R, at the end (f) of which the wire is burned. The concave mirror O reflects the light in parallel rays.

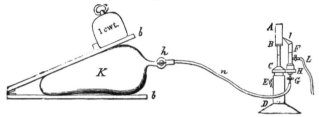

Fig. 19.

By means of the handle H, the lamp with parallel rays can be turned in any direction, and the clockwork can be instantly stopped by the key m.

The magnesium light is surpassed in power by Drummond's lime light. This is produced by a gas or spirit flame, into which oxygen gas is blown. The oxygen gas is produced by the action of heat on chlorate of potassium, a salt rich in that element. On heating this salt, oxygen is given off as a gas, and is collected in an india-rubber bag K. (See Fig. 19.)

This bag is closed by means of a stopcock h, and when

used is placed between two boards *b b*, a weight being placed on the upper one. By the pressure of this weight the oxygen gas is forced through the cock *h*, and the india-rubber pipe *n*, into the lamp *D*. To this is attached a burner *H F*, terminating in a jet *I*. The coal gas which serves for combustion enters through the cock *L*, which is connected with a gas tube.

The combustion takes place at the jet *I*. Without oxygen the coal gas burns with a bright soot-producing flame; but as soon as the oxygen is turned on, the flame becomes smaller and blue in colour, and burns with an intense heat.

Its illuminating power is small, but as soon as the flame has sufficiently heated the lime cylinder *A B*, a dazzling white light is produced, which has a very intense effect in photography, and has been used with success by Monckhoven and Harnecker for enlarging photographs.

Fig. 20.

The same apparatus serves for the production of what are called dissolving views.

The electric light, produced by help of a galvanic battery, has a still more powerful effect than the lime light.

If a piece of gas carbon *k* (Fig. 20) and a piece of zinc are dipped together into an acid (dilute nitric or sulphuric acid), electricity is developed, which produces a spark on bringing together the two ends of the zinc and coal above the fluid; this spark is, however, very feeble. But if several vessels containing zinc cylinders and pieces of carbon are employed, the spark becomes very intense; and, as we are able to increase to any extent the number of these elements, we are able to produce a cone of light of any degree of brilliancy, exceeding all other artificial light.

In arranging electric batteries of this kind, the zinc of one element is connected with the carbon of the second, and the zinc of this with the carbon of the third element. (See Fig. 22.)

If the two wires from Z and C are brought together, a spark of light is produced by the electric stream burning the wire.

The light is generally produced between cones of carbon placed in front of a concave

Fig. 21.

mirror H (Fig. 23). SS is a self-acting apparatus to regulate distance between the carbon points KK. The

Fig. 22.

upper point is connected with the wire K by the stand F, the lower one with the wire Z of the electric battery. Thirty-six cells similar to those of Fig. 21 suffice to produce the electric light.

The preparation of the battery makes the application

of the light inconvenient. In other respects this light
surpasses all others in photographic effect.

Nadar has made with it many excellent pictures of
the catacombs of Paris. It has also been used to take

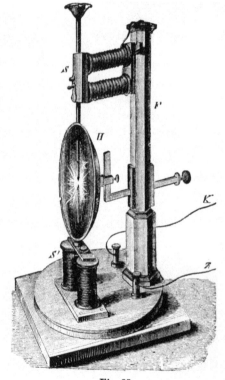

Fig. 23.

portraits. But the employment of such a dazzling artificial light is attended with the drawback of occasioning
harshly defined shadows, which disfigure the portrait.

This difficulty has been overcome by placing another electric light of less power on the shaded side; but it is difficult under this dazzling light, as in direct sunlight, to prevent contraction of the features.

It thus appears that all these artificial lights can only serve as auxiliaries for photographic purposes, for they are, moreover, very expensive. Accordingly, their use will be confined to places that cannot be lighted in any other way. The writer has used sunlight with great advantage in taking photographs of Egyptian sepulchres. He

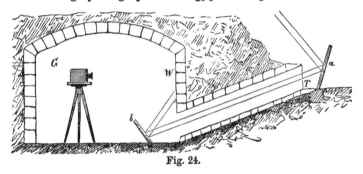

Fig. 24.

brought the light through subterranean passages by means of reflection.

Let the reader imagine a mirror a set up in the open air, reflecting the sun's rays through the entrance T (Fig. 24), into the subterranean vault G. In this vault they are received by a second mirror b, which throws the rays on the surface of wall W, of which a photograph has to be taken. It is true that nothing but a spot of light is thus produced; but if, during the exposure of the photographic plate, this spot be moved over the part of wall W,

of which a photograph is to be taken, all parts of the
object receive successively enough light to produce photographic effects. The movement of the spot of light over
the wall is effected by moving the mirror b.

Braun of Dornach, by help of the same method, was
able at a later date to reproduce the very dark frescoes
of Raphael and Michael Angelo in the Sistine Chapel
and in the galleries of the Vatican, with excellent results.

Sunlight remains the most important source of light
for photographic purposes. The brightness of this light
is, however, subject to great variations. Even the eye
recognizes that the sun is much brighter at noon than in
the morning and evening. According to the measurements of Bouguer, this difference is so considerable, that
the sun at an elevation of 50° above the horizon is 1200
times brighter than at sunrise. The eye, moreover, perceives a decided difference of colour between the sun on
the horizon and the sun at the zenith. The latter appears
white, the former of a reddish hue ; and, on examination
with the spectroscope, it is found that in the setting sun
the reddish rays predominate, while the blue and violet
are in part wanting.

For this reason the chemical action of sunlight is very
feeble in the morning and the evening; it increases as the
sun rises above the horizon, and it attains its greatest
intensity about noon.

The cause of the red hue of the morning and evening
sun is found in the fact that the particles of the air partly
reflect the blue rays—for which reason the air (that is, the
sky) appears blue—whereas they transmit the yellow and
red rays more easily.

If E (Fig. 25) is the earth surrounded by the atmosphere A, S the sun at the moment of sunrise, S'' the sun at the moment of sunset for the place O, and S' the sun at noon, it is apparent that the sun's rays at sunrise and sunset have to travel much further through the atmosphere—namely, the distance between a and O—than when the sun is in the zenith S'. But in proportion as the stratum of atmosphere through which the sun must pass to arrive at the spectator is thicker, the weaker becomes the light. It follows from this that on high mountains the chemical action of the rays of light must

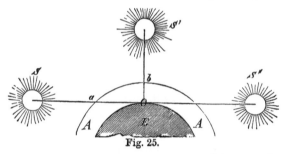

Fig. 25.

be more intense, and this has been proved by experiments on the Alps to be the case.

But not only are chemical effects produced by direct sunlight; the light of the blue sky, which is nothing but reflected sunlight, is likewise chemically active, and powerfully so, through its blue colour.

It has been already stated that the blue colour of the sky proceeds from the fact that the particles of the air reflect blue light. The quantity of this reflected light varies with the hour of the day, being strongest when the sun is highest (that is, at noon), and diminishes in proportion as the sun approaches the horizon. Photographers therefore prefer the middle of the day, *i.e.*, between 10 a.m. and 2 p.m., for taking portraits, for which the light of the blue sky only is used. During these hours the chemical effect of this light remains almost the same; afterwards it begins to diminish—quickly in winter, more slowly in summer. Thus the chemical power of light, according to Bunsen, expressed in degrees, is at Berlin:—

	12 o'clock.	1 o'clock.	2 o'clock.	3 o'clock.	4 o'clock.	5 o'clock.	6 o'clock.	7 o'clock.	8 o'clock.
June 21	38°	38	38	37	35	30	24	14	6
Dec. 21	20°	18	15	9	0	0			

It appears from this example how extraordinarily weak is the chemical action in winter (thus about noon on the 21st December only half as powerful as at noon on the 21st June); moreover, how small is the amount of chemical light which is diffused by the blue sky on the 21st December, on account of the shortness of the day. Therefore a longer exposure is necessary in winter than in summer, and, the printing process being slower, a longer time is required in winter to print the same number of pictures.

Now, the intensity of the blue sky light depends on the position of the sun, and the latter varies, not only according to the different seasons, but also at the very same seasons on different parts of the earth.

If circles be drawn round the earth from pole to pole, we obtain what are called meridians (*m m* Fig. 26). At all places situated on the same meridian it is noon at the same time, but the height of the sun varies very much according to the distance of the place from the equator.

If circles be drawn round the earth parallel to the equator, they form the so-called lines of latitude. If the sun is perpendicular at noon at a particular place on the equator, at 10° of north latitude it is 10° lower; that is, the height of the sun (or the distance of the sun from the horizon expressed in angular measurement) is 80° At 10° further north, the position of the sun, at the same time, is only 70°; and at the pole, which is 90° from the equator, the height of the sun $= 0$; that is, the sun is on the horizon.

Fig. 26.

The chemical action of the blue sky light varies greatly, corresponding to the different positions of the sun at the same time. Thus, for example—

At Cairo, on the 21st Sept., the strength of light at noon $= 105°$
At Heidelberg   „   „   „ $= 57°$
In Iceland   „   „   „ $= 27°$

Therefore, the more southerly a place is, the richer it is in the amount of light available to the photographer. Accordingly, the American photographers are better off than those of Germany and England.

These differences in the chemical intensity of light are also essentially modified by the state of the weather. If the sky is covered with grey clouds, the chemical intensity of the light is considerably less than with a perfectly clear sky. On the other hand, white clouds increase the power of the light very decidedly. In the autumn the chemical intensity of light is much greater than in spring, perhaps in consequence of the greater transparency of the air. According to Roscoe, it is in August and Sep-

tember more than one and a half times as great as in
March or April.

These variations in the chemical intensity of light are
very important to the life of plants. The green leaves
of plants inhale carbonic acid and exhale oxygen under
the influence of light. This breathing process does not
take place without the presence of light. The green
colour of leaves and the variegated colours of flowers
only exist under the operation of light. In the dark,
plants only develop sickly blossoms, like the well-known
white sprouts of potatoes kept in cellars.

The necessity of light for the life of plants is also seen
in the effort made by plants kept in darkened rooms to
reach the apertures which admit light. Hence a plant
develops with an energy proportional to the intensity of
the light. The greater fruitfulness of the tropics is to
be ascribed, not only to the higher temperature, but also
to the greater chemical intensity of the light. Recent
observations have established that the yellow and red
rays, and not the blue and violet, produce the greatest
chemical effect on the leaves of plants.

We have now arrived at the knowledge of the importance of light for the economy of nature. Atmospheric
air is a mixture of two gases, oxygen and nitrogen.
Nitrogen is a perfectly innocuous kind of air, serving to
dilute the oxygen; for the latter, though essential to life,
is injurious if undiluted.

In breathing, part of the oxygen is absorbed in the
lungs: it forms, with the organic constituent parts of the
body, carbonic acid and water. The carbonic acid and
water are exhaled by us and dispersed again in the air.

It is easy to prove by an experiment that a considerable amount of carbonic acid is contained in the air we exhale. Carbonic acid forms with lime-water an insoluble precipitate, carbonate of lime. If now we blow the exhaled air through a glass tube into perfectly clear lime-water, the latter becomes milky by the formation of carbonate of lime. Hence, by breathing, the amount of oxygen in the air is continually diminished and converted into carbonic acid. The same result is produced on a larger scale by the process of combustion. In this process a combination of wood or coal with oxygen takes place, and the result is again, principally, carbonic acid.

It might be supposed from this fact, that, in the course of time, the amount of oxygen in the air must diminish, while that of carbonic acid would increase. This actually takes place in closed spaces. Leblanc found that, after a lecture in one of the lecture-rooms of the Sorbonne at Paris, the air had lost one per cent. of its oxygen.

In the open air no such a diminution of oxygen and increase of carbonic acid gas can be detected, and the reason of this is that the carbonic acid formed by combustion and the exhalations of animals is again decomposed by plants under the influence of light.

Plants absorb the carbonic acid, retaining the carbon and liberating the oxygen; by which means the latter, lost by combustion and exhalation, is made again available.

There was a time when the atmosphere was much richer in carbonic acid gas than now. When the incandescent and fluid masses that once formed our earth gradually solidified, when the aqueous vapours were condensed as seas, the atmosphere contained almost all the carbon of the earth, combined with oxygen, as carbonic acid gas. The air was therefore at that time infinitely richer in carbonic acid than now. When at length the earth had cooled sufficiently for vegetation to be developed, gigantic plants shot forth from the warm ground under the influence of the sunlight. They flourished luxuriantly in an atmosphere so rich in carbonic acid, the carbon of the carbonic acid passed over into the form of wood, and thus for thousands of years the carbonic acid in the atmosphere was continuously diminished. Revolutions of the earth's surface succeeded; whole tracts were buried under sand and clay; their forests decomposed, and were changed into coal. A fresh vegetation sprouted forth from the newly formed soil, and again absorbed, under the influence of light, the carbonic acid of the atmosphere, to be once more engulfed. Thus the carbon from the carbonic acid of the atmosphere was stored as coal in the depths of the earth; and thus the atmosphere, by the chemical effect of light, became continually richer in oxygen, until at length, after countless ages, it attained that wealth of oxygen which made the existence of man possible.

We see, therefore, that the chemical influence of light has played an important part in the development of our planet, and that it continues to do so in the economy of nature.

CHAPTER IX.

ON THE REFRACTION OF LIGHT.

Simple Refraction—Index of Refraction—Refraction in Glass Plates—
Prisms and Lenses—Production of Images by Lenses.

WE have already pointed out (p. 60) that when a ray of
light passes the border of two transparent media of unequal density, a change of direction
takes place which is called refraction.

If a small coin is placed in an
opaque cup, and the eye be kept in
such a position that the edge of the
vessel just conceals the coin, it becomes
visible on pouring water into the cup,
and this takes place by the refraction which the rays
experience in passing from the water to the air. (See
Fig. 27.)

Fig. 27.

The angle which the rays make, before and after the
refraction, is called the deflection.

This deflection increases in proportion to the oblique-
ness with which the rays fall upon the surface of the
water.

In order to determine exactly the degree of the refraction, let a line be conceived to be drawn at right angle

G

to the surface through the point of immersion n of the
ray $n\,l$ (Fig. 28). This line is called the normal, and the
angle i which the ray forms with this normal is called the

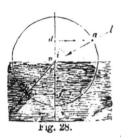

Fig. 28.

angle of incidence, while the angle r
which the refracted ray forms with
the same normal is called the angle
of refraction.

The ratio of the magnitude of
the angle of incidence to the angle
of refraction is peculiar. If a circle
be described with centre n, and from
the points a and b perpendicular
lines $a\,d$ and $b\,f$ are let fall on the
normal, these lines are what mathematicians call the sines
of the angles. Thus $a\,d$ is the sine of i, and $b\,f$ the sine
of r. The ratio of the sine of the angle of incidence to
the sine of the angle of refraction is constant.

This ratio is, when light leaves air for water, 4 to 3;
that is, the sine $b\,f$ is $\frac{3}{4}$ times as great as the sine $a\,d$, or
the sine $a\,d$ is $\frac{4}{3}$ times as great as $b\,f$. Light is still more

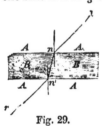

Fig. 29.

refracted on entering glass. In this
case the ratio of the sines is as 3 to
2. This ratio of the sines of the two
angles is called the index of refraction.

If a ray of light $n\,l$ falls upon a
smooth sheet of glass (Fig. 29), it experiences a similar refraction; it con-
tinues in the direction $n\,n$, and the sine of the angle of
refraction at n in the glass becomes two-thirds of the
sine of the angle of incidence.

On issuing from the other side of the sheet of glass,

another refraction takes place; but in this case the sine
of the angle of refraction at n' in the air becomes one and
a half times that of the angle in the glass, and as the
angle at n is equal to the angle at n', the angle of emer-
gence of r n' is of the same magnitude as the angle of incidence of n l; that is, the ray continues, after refraction,
in its original direction.   At all events, it only experiences
a shifting parallel with itself.   Therefore we see objects
through our windows in the same direction in which they
are really situated.

The result is entirely different
when the spectator looks through
a triangular glass (Fig. 30).   If
the eye is at o, and an object at
a, and a triangular prism be held

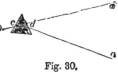

Fig. 30.

close to the eye, the object does not appear to be at
a, but in the direction of a'.   The incident ray a d suffers
a deflection at the first face of the glass, taking the direction d c; at the refraction on the second face it takes
another, o c.   Both deflections are in the same direction.

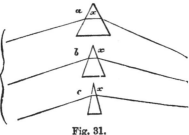

Fig. 31.

The greater the magnitude of the angle x which the
two faces of the prism, through which the ray passes,
make with each other, the greater is this deflection.   Thus

the deflection by the prism b (Fig. 31) is greater than by the prism c, and by the prism a it is greater than by b; because the angle of refraction x is greater in b than in c, and in a it is greater than in b.

If a glass structure be erected, consisting of separate prisms of varying angles (Fig. 32), and if a bundle of parallel rays be conceived to fall upon it, the ray a will

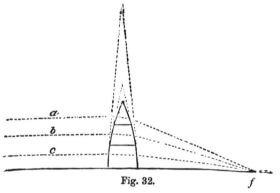

Fig. 32.

be more strongly deflected than b, which falls on a prism having a smaller angle; and the latter, again, will be more deflected than ray c, and the result may be that all the rays unite in one point f.

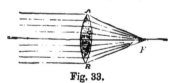

Fig. 33.

If instead of the separate prisms we substitute a solid symmetrical mass of glass, we obtain the section of a burning glass, or, as the opticians say, a lens, which has the property of uniting all parallel incident rays in one point. (See Fig. 33.)

Every lens is contained between two curved faces.
The connecting line running through the centre of the
two surfaces is named the axis of the lens, and the point
E (Fig. 33), where the parallel incident rays unite, is the
focus, while its distance from the lens is the focal length.
But not only are parallel rays united in one point by the
refraction of a lens of this kind, the same thing occurs
with the divergent rays which issue from any luminous
point. The point in which such rays are united is called
the conjugate focus of the luminous point.

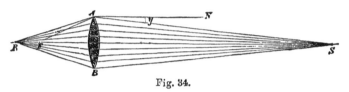

Fig. 34.

A luminous point *S*, for example, sends a cone of rays
to the lens. After refraction these are united at *R*. If *S*
be brought near to the lens, *R* removes further; if *S* be
brought so near that its distance from the lens is twice
the focal length, then the converging point *R* is equally
distant from the lens.

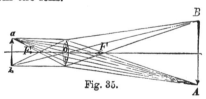

Fig. 35.

If instead of the luminous point an object (for example
an arrow *B A*) is placed before the lens, from each
individual point of the object a cone of light proceeds to

the lens, and all the rays of one and the same cone converge in one point, the rays from A in a, and those issuing
from B in b; and the result is that a perfect miniature
and inverted image of the arrow is produced.

If the arrow be moved nearer to the lens, its image is
removed farther from the lens and becomes larger. For
example, if the little arrow a b is placed before the lens,
it produces the enlarged picture A B.

But if the arrow be removed farther from the lens, its
image approaches the lens, and becomes continually
smaller. Accordingly, a lens is able to project enlarged
or diminished images of an object, the size of the image
varying with the distance of the object from the lens.

CHAPTER X.

PHOTOGRAPHIC OPTICAL APPARATUS.

Construction of the Camera Obscura—Telescopic Images—The Magic
Lantern—Enlarging Apparatus—The Stereoscope.

WE have just shown that a lens is able to produce
enlarged or diminished pictures of objects according to
their distance. On this depends the property of the

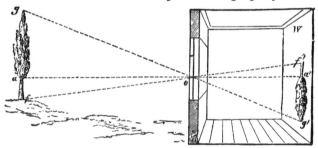

Fig. 26.

camera obscura, the most important photographic instrument, to project plane pictures of solid objects.
We have already described (see p. 7) its simplest
form. It is a dark chamber having a small hole in the

shutter. This arrangement produces indistinct and weak pictures. But if a lens is placed in the hole of the shutter *o* (Fig. 36), a brighter and more distinct image is cast on the wall than that of a simple hole. It is evident that in this case the distance of the wall from the lens must correspond to the distance of the object. Now, as this varies, the dark chamber has been converted into a small box (Fig. 37), the back part of which is movable, and

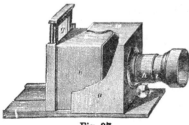

Fig. 37.

contains a ground-glass slide *g*. If the back of the camera *o* is moved to and fro, it is easy to find the exact situation of the image of an object placed before the lens *l*. In order to determine this distance with the necessary accuracy, photographic lenses are supplied with a rackwork motion *r* in the frame of the lens; but this addition is by no means necessary.

In order to be able to see the image on the ground-glass slide *g*, all foreign light which would blind the eye must be kept off, and for this a dark cloth is thrown over the head, forming what is called the focussing cloth.

The operation of seeking the image is named in photography focussing. It follows from what has been said that the image appears inverted on the ground-glass slide. Though the process of focussing appears simple at first sight, it is really rendered difficult by the fact that objects at different distances from the lens form images that likewise vary in distance from the ground-glass slide. For example, if a head be placed opposite a camera, the nose is nearer to the lens than the hairs behind the head; and the result is that the image of the nose in the camera is farther from the lens than the hairs of the back or side of the head. Accordingly, the whole head never presents the same degree of sharpness or definite accuracy. Photographers are satisfied with obtaining the main points with definite clearness, such as the face, and they bestow less care on the subordinate parts.

If the situation of the object be rather remote (for example, a landscape, whose nearest features in the foreground are distant about fifty times the focal length), the images of the various objects, whatever their distance may be, appear all in the same focus.

The same thing occurs in the case of stars. Photographic cameras are well adapted to project images of stars, only they are very small if the focus of the lens is short. Accordingly, telescopic lenses are preferred in such cases. The production of images in this case depends on the same principle as the production of images by other lenses. If we imagine a telescopic lens $o\ o$ (Fig. 38), and an arrow $A\ B$ placed before it at a great distance, a very diminished image $b\ a$ would be formed. Thus, the image of

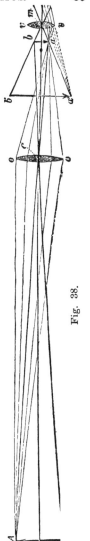

Fig. 38.

the sun at a distance of 90 millions of miles, when cast by a lens of six feet focal length, is only eight lines in diameter. If it is wished to obtain the photograph of such an image, the tube R, on which the lens L is fixed, must be arranged as a photographic camera. (See Fig. 39.) A movable ground-glass screen n is used to focus the image, to be exchanged for a photographic plate during the exposure. This is the method adopted by Warren de la Rue, Rutherford, and others engaged in eclipse expeditions, and also by the author in the expedition to Aden in 1868.

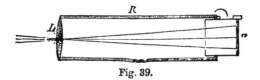

Fig. 39.

The images taken by a photographer are usually smaller than nature. But he is also able to produce images that are larger than the originals. Every lens gives, as described at p. 86, images of the same object, varying in size according to distance. If the object is nearer than twice the focal distance, an enlarged image is produced, but if it is more remote, the image is smaller. The latter is the commoner case. The production of enlarged images direct from nature is attended with difficulties. The larger the image, the larger is the surface over which the light from the objects is spread, and, accordingly, the smaller will be the amount of light over each part of the image. But, in proportion as an image is deficient in light, the longer the exposure must be to produce a photographic impression. A man would find it difficult to endure such a long sitting, therefore this method is only employed for drawings and the like.

Enlarged pictures of other objects are produced by the help of an apparatus resembling the magic lantern. The magic lantern serves for the production of enlarged images by means of lenses. Instead of a simple lens, a system of lenses *n n o o* (Fig. 40) is employed for enlarging, which gives more sharply defined images. The object is painted or photographed on glass plates, which are placed in the slide *a a*, and brightly illumined by the lamp *L*. The concave mirror *H* and the lens *m m* are employed to concentrate the lamplight on the object that has to be enlarged. The images obtained

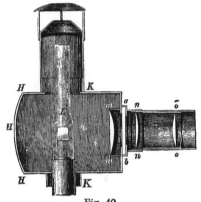

Fig. 40.

vary in size with the distance of the screen from the lens.

This instrument was formerly nothing but a plaything, but it has latterly become an important auxiliary in instruction. Photographs of microscopic preparations, of animals, plants, minerals, landscapes, national types and architecture, may be in this manner more faithfully represented than by maps, which are in general very imperfectly designed.

In America this application of the magic lantern is universal. Every educational establishment of any size possesses one such instrument, and often more. In Germany, though so useful, it has been hitherto left in the hands of showmen connected with the annual fairs, who employ them for what are called dissolving views. These dissolving views are produced by the help of two magic lanterns placed side by side, both of which project their images on the same screen. If one of these lenses be covered up, one of the pictures disappears and the other alone remains visible. Meanwhile, if another picture be substituted for the one withdrawn, and the lid again taken off, a combination of two images is obtained. If the one lens be only gradually, not suddenly, closed, the corresponding image also fades gradually away.

Professor Czarmak, at Leipsic, has latterly introduced the representation of enlarged images by the magic lantern as a useful auxiliary in his lectures; and he obtained so great a success with it, that he prepared the way for its general introduction into schools.

We take this occasion to remark that wonderful pictures on glass have been lately produced from photographs by means of a new printing process. These glass pictures are now for sale, and are specially designed for the magic lantern. Their price is so moderate, and the objects they represent are so interesting—landscapes from all parts of the earth—that it is within the reach of every family to obtain a collection of the most beautiful and interesting views. At domestic entertainments by the family fireside, such pictures united with a magic lantern become an important means of instruction and enjoyment to both young and old.

A petroleum lamp is not sufficient for the representation of such images on a large scale. For this purpose more powerful sources of light must be employed, such as the lime light or the electric light (see p. 72). To obtain photographs of such enlarged images, a sheet of sensitized paper is stretched over the place and instead of the screen.

To produce life-size photographs, the magic lantern is not used, but the so-called solar camera, a section of which is represented in Fig. 41, and a view of its exterior Fig. 42.

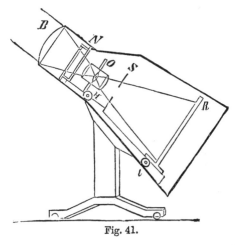

Fig. 41.

Sunlight is made to fall on a large lens B, which concentrates it on the small negative N; close to which is the objective O, which projects an enlarged image on the screen R. The image will be obviously a negative one. If a sensitized piece of paper be stretched at R, this paper becomes brown at all places where the negative is clear (transparent), and it remains white in all places where the negative is black (opaque); therefore the resulting picture will be positive. The whole apparatus is enclosed in a light-tight wooden box, which can be shifted by means of a rackwork motion, so that it can always be turned towards the sun.

In conclusion, we have to describe a most beautiful

Fig. 42.

optic-photographic apparatus, which enables us to see images not only as plane objects, but as solid bodies. This is the stereoscope.

Our readers know already that this instrument is intended to exhibit double pictures, the two halves of

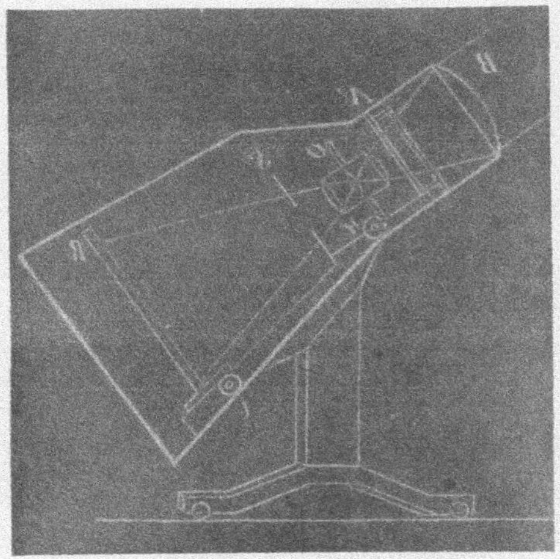

PLATE II.

Negative.

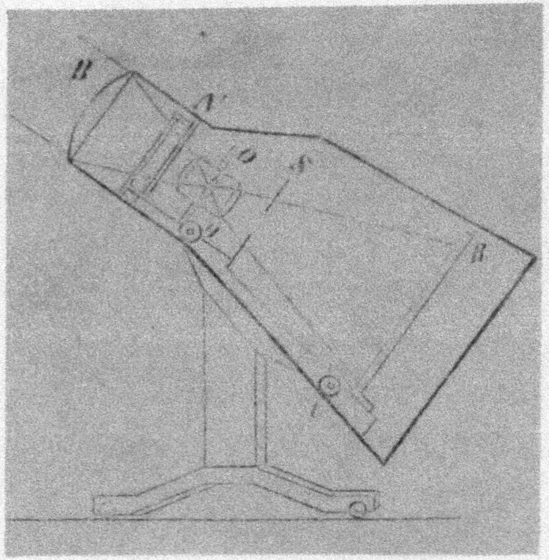

Positive.

NEW TALBOTTYPE OR LICHTPAUS PROCESS.

which at the first glance seem to be absolutely alike, and
which when viewed through the instrument form one
picture, which appears no longer plane but solid.

The two pictures which are seemingly alike are, in
fact, different. If we look at a cube with the right eye,
we see rather more of the right side; if we look with the
left eye, we see something more of the left side, taking
for granted that the head is not moved. The pictures of
the right and of the left eyes are combined with each
other and give the impression of solidity.

If we close one eye, the impression of solidity is far
weaker; the objects appear plane or flat. This may not
be readily credited, because men do not often seek an
explanation of what they see, but look at objects much
too hastily. But it can be easily ascertained that such
is the fact if a bottle be placed before a wall, or in front
of an upright book. If we then look at them with both
eyes, we readily perceive the distance of the bottle from
the wall or book, but directly we close one eye the bottle
and the book appear to be almost contiguous, and it is
only by moving the head on one side that we clearly
distinguish the distance between them.

Accordingly, the use of both eyes is necessary for a
perception of solidity. It is only in this manner that we
come to the conviction that space has not only height
and breadth, but also depth. One-eyed persons only
receive this impression by turning the head on one side.
If objects are very remote, the difference between the
views which the right eye and the left eye have of them
is very inconsiderable; and, accordingly, such remote
objects appear flat and without solidity, and it is only
when we change our position and observe them from

different sides that we become acquainted with their
solidity. This is therefore purely an affair of experience.
Every person will recognize a distant house as a solid
object, because we know from experience that a house is
solid; but that we actually see it as a flat surface is proved
by the deception produced by theatrical decorations,
where the remote background, if properly painted, often
produces an extremely natural effect; but we perceive
this background to be flat directly we move the head
on one side. When this is done, a solid object presents
a different appearance, but a flat surface remains unchanged.

Wheatstone was impressed by the fact that the solid
impression made by an object is caused by the combination of the different views of it by the right and left eye.
Accordingly, he tried to substitute for a single picture a
view of the right side of an object for the right eye, and
of the left side for the left eye. He obtained in this
manner a perfectly solid impression, though the double

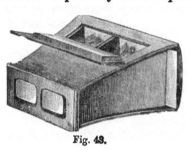

picture occasioning it is
no solid at all. Some
people indeed are able to
see stereoscopic pictures
as solids without using
an instrument. But
most persons require an
apparatus which renders
it possible for both eyes
to see in the same place

Fig. 43.

the two separate pictures. This apparatus is the stereoscope (Fig. 43). Its most essential features are, the two
pictures on the slide and the partition in the interior of

the box which prevents the right eye seeing the picture on the left, and vice versâ; and further, the lid, which is generally provided with a mirror which can be either shut or opened, so as to exclude or let the light into the box; and, lastly, the two lenses at the top.

These lenses are represented in the diagram (Fig. 44); they are two halves of one lens, and work in the same manner. We are indebted to Brewster for the construction of this instrument.

Fig. 44.

We have shown at a previous page that a lens gives a diminished image of a remote object, and an enlarged inverted image of a near object. This image is objective, that is, it can be rendered visible on the ground glass screen of a camera. Nevertheless, this phenomenon only takes place when the object is more remote than the focus. The case is different when the object is nearer to the lens. Let an ordinary magnifying or burning-glass be held near some writing, and it will be seen upright, and not inverted. The image appears also enlarged, but on the same side as the object, and the accompanying diagram illustrates the manner in which it

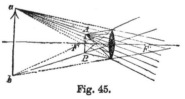

Fig. 45.

originates. Thus F represents the focus of the lens, $A\ B$ an object within the focal distance, and $a\ b$ its image, as it appears to the eye upon the other side of the lens. As may be seen in the diagram, the rays issuing from $A\ B$ do not actually unite to form an image, but their directions

H

prolonged through the lens (the dotted lines of the figure), unite to form the image *a b*, and there the magnified image will be seen. The eye seeks the objects seen in the direction of the rays which fall upon it, as may be seen, for example, in a mirror, where we see the mirrored objects behind.

In order to give a clear illustration of the different way in which a lens represents near and distant objects, we introduce the diagram of p. 85 again, which shows the production of an inverted and enlarged image *B A* representing the arrow *a b* situated without the focus.

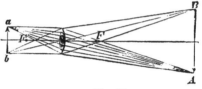

Fig. 46.

A lens employed to see objects enlarged within the focus is named a microscope, or magnifying glass. The lenses of a stereoscope are such magnifying glasses. They present us with a rather enlarged upright image of the object seen, but they produce likewise the effect of a prism. As may be seen from Fig. 44, the two lenses consist properly of only two half lenses, which are placed in a reversed position.

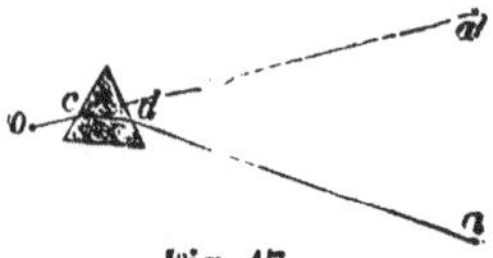

We pointed out at a previous page that an eye *o* sees an object *a* through a prism in the direction *o a'*, that is, shifted towards the angle of the prism. The same thing happens with stereoscopic glasses. We see

Fig. 47.

the image, not in the original direction, but reflected towards the angle of refraction, that is, towards the centre of the instrument.

The two corresponding points $a\ a\prime$, which belong to the right and left images, appear therefore in common to both eyes at $a\prime\prime$—that is, at the same place —and consequently our two eyes see only one image instead of two.

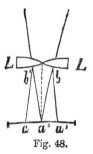

Fig. 48.

Now, every one who wishes to see an object (for instance, writing) clearly and distinctly holds it at a definite distance from his eye. This distance is the distance of clear vision. In the case of good eyesight it is eight inches; with far-sighted persons it is more, and with short-sighted persons less. In the case of stereoscopic vision, the image appears remoter or nearer, according as it is removed from or approached towards the two lenses. If the image is near the lenses, it appears when viewed through them nearer and smaller. In the opposite case it appears farther and larger. But every one wishes to see the image at the distance of clear vision, therefore stereoscopes must have movable lenses, in order that persons may adapt the position of the object to the eye; that is, that they may vary the distance of object and lens until the image appears clearest. If such focussing arrangements are absent, the instrument is only adapted for eyes of average power of vision, and requires an effort in eyes of a different calibre. Persons are often met with whose eyes are not of equal strength, one being short and the other long-sighted. There can be no satisfactory stereoscope for such persons; for if the distance of the lenses is adapted for one eye, it does not suit the other.

Nevertheless, such persons can obtain a tolerable stereoscopic effect if they hold a suitable eyeglass before one eye.

A great hindrance to viewing stereoscopic pictures on paper is the shape of the Brewster stereoscopic box, which is closed all round and only open at the top. This aperture only admits an insufficient amount of light to the picture, which is commonly left in the shade on one side.

This defect has been removed in the American stereoscope, which dispenses with any box. Lenses are fitted in a frame $g\ g$ (Fig. 49), which may be held by a handle;

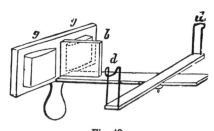

Fig. 49.

the partition b serves to separate the field of view of the two lenses. The object is placed on the cross-board $d\ d$, and this board can be easily slid to and fro, so that the proper position of the image with reference to the eye may be found.

But the American stereoscope is only suitable for paper pictures. The beautiful transparent stereoscope pictures on glass can only be viewed with Brewster's stereoscope, as they must be seen by transmitted light, and all front light excluded, or the effect will be destroyed.

We have named the stereoscope an optic-photographic apparatus. But we remark that double pictures drawn by the hand can also be viewed through it. It is evident that the preparation of such pictures only succeeds in very simple subjects. It would be very difficult to represent stereoscopically a complicated object; for example, a man, a landscape, or a machine. Photography, which can produce with the greatest ease pictures of the most complicated objects from any point preferred, rendered this possible. It is only since the invention of photography that the stereoscope, which was formerly a philosophical instrument, has become a favourite instrument with the public. Notwithstanding their small form, the pictures of these instruments make a clearer and more intelligible impression than single pictures of the same object in a larger form. A single picture of a machine, or of complicated architecture (for example, the choir of the Cologne Cathedral), is often a hopeless maze of details. But in the stereoscope the confused masses are directly defined; they become distinct in perspective, and the eye perceives with great clearness the interior structure. In this respect the stereoscopic pictures are of equal value to the magic lantern in imparting instruction.

CHAPTER XI.

THE CHEMICAL EFFECTS OF LIGHT.

Physical and Chemical Processes—Moser's Experiments—Action of Light upon the Elements—Phosphorus, Oxygen, and Chlorine—Action of Light upon Salts of Silver—Chloride, Bromide, and Iodide of Silver —Theory of the Developing Process—Dry Plates—Theory of the Positive Process.

IN the previous chapter we have become acquainted with the part which light plays in the processes of photography. We will now enter the domain of chemistry, in order to explain the phenomena which occur when substances sensitive to light are exposed to its action.

All bodies in nature are perpetually subject to change. The sun, moon, and stars change place; wood and sugar can be powdered; lead can be melted, and thus its state of aggregation altered. In such cases the matter is not changed. Wood may be rubbed or sawn into the finest dust, yet it remains wood; lead remains lead, notwithstanding the melting. Changes of this kind, that leave the matter of bodies unchanged, are styled physical changes.

But, beyond the physical changes, there exist others of a different nature. If a piece of wood be heated in flame, it burns and loses its character as wood. It becomes changed partly into combustible gases, partly into ashes. A rod of iron, heated to redness in the air, becomes dull and coated with a black crust, which falls off in flakes when struck by a hammer. In this case the substance of the iron is totally changed. Changes of this kind are styled chemical changes.

Now, light is able to produce both chemical and physical changes. We have already stated that the red mineral, realgar, falls into a yellow powder when exposed to light. This is a physical change, for the yellow powder is still realgar, and if it is fused it forms, on cooling, compact red masses, which are again changed on exposure to the light. The number of physical changes of this kind occasioned by light is not great, but the phenomena are in themselves remarkable.

Moser has remarked that light has a certain action on almost all surfaces. He covered smoothly polished surfaces of silver, ivory, and glass with perforated screens, and exposed them to the light. After this he breathed upon them, or exposed them to the action of the vapour of mercury, and found that the vapour was condensed most powerfully where the light had reached the surface. Accordingly, Moser established the proposition : Light reacts on all bodies, and its action can be made visible by the greater condensation of vapours on the parts exposed to light.

The chemical changes effected by light are far more numerous than the physical, and their study is the special province of photo-chemistry.

Before passing to the more complicated phenomena of photography, we must make the reader acquainted with the simpler phenomena of the action of light.

(a) *Action of Light on the Elements.*

The chemist understands by the term elements simple undecomposable bodies. Thus water, which the ancients named an element, is no element in the chemical sense of the term, for it can be easily decomposed into two components of a gaseous nature—oxygen and hydrogen. Air, also an element of the ancients, is no element viewed in the light of chemistry, for it is a mixture of two gases—oxygen and nitrogen. But these two substances, oxygen and nitrogen, are undecomposable bodies, or elements. The chemical elements contain amongst their number all the known metals, also sulphur, phosphorus, chlorine (a greenish, strong-smelling gas) ; further, the less known substance bromine (a brown, unpleasantly smelling fluid) and iodine (a black, volatile solid). Each of these elements can combine with others to produce bodies with new properties. Metallic iron combines with the gaseous oxygen, and produces the red iron rust. Sulphur unites with oxygen, and produces the pungent, strong-smelling sulphurous acid. Iodine and chlorine combine with metals forming the iodides and chlorides, amongst which are iodide and chloride of silver.

It is remarkable that many elements present themselves in quite different forms, so that it might be supposed they were different substances. The yellow, inflammable, poisonous phosphorus, soluble in ether, and formerly used in the manufacture of matches, is changed by heating in a closed vessel into a reddish substance difficult to kindle, not poisonous, and insoluble. This is, however, phosphorus, and passes by melting into the state of common phosphorus.

It is an interesting fact that this transformation of yellow into red phosphorus is effected not only by heat, but also by light. If yellow phosphorus be exposed for a long time to the light, it becomes red.

The oxygen of the air is also susceptible of similar changes. Ordinary oxygen is a colourless and inodorous gas. By the action of electricity, however, it is easily changed into another kind of gas, distinguished by a peculiar smell—the so-called sulphurous smell of lightning. This new gas, ozone, has a much more oxidizing or rusting effect than common oxygen.

Ozone is also formed by the action of light. If oil of turpentine be poured into a large bottle containing air, and agitated violently in the sunlight, ozone is formed.

Equally peculiar are the changes experienced in sunlight by two other elements not so well known, chlorine and bromine, which have only been carefully observed latterly.

Chlorine is a yellowish-green gas, with a disagreeable smell, distinguished by its properties of bleaching coloured stuffs and destroying infectious matter. Bromine is a substance very similar to chlorine, but in a fluid not in a gaseous state at ordinary temperatures, though it can be easily vaporized, and then appears as a brownish-red gas.

Both chlorine and bromine gas show a peculiar relation to light, even in combination.

Chlorine gas behaves in a peculiar manner with hydrogen—a gas which forms one of the constituents of water, from which it can be easily obtained by the action of zinc and sulphuric acid. The zinc attracts the oxygen of the water, forming, with the sulphuric acid, sulphate of zinc, while the hydrogen escapes in the form of gas.

If this combustible gas is mixed with chlorine, and the mixture is exposed to the sunlight, an explosion takes place. This accompanies the chemical combination of chlorine and hydrogen to form a new body—*hydrochloric acid*—having no resemblance to chlorine or to hydrogen. This acid is of a sour taste, very soluble in water, does not bleach like chlorine, and is not combustible.

Another body—iodine—is very closely related to chlorine and bromine. It is a solid, appearing in the form of shining black crystals, and giving when heated a wonderful violet vapour.

(b) *Chemical Action of Light on Salts of Silver.*

Iodine, bromine, and chlorine unite with metals, forming the iodides, bromides, and chlorides of the metals. Kitchen salt is one of the commonest combinations of this kind, consisting of chlorine and sodium. Sodium is a metal not employed in the industrial arts, which possesses the property of powerfully attracting the oxygen of the air, or rusting, so that it has to be protected by being kept under naphtha. The chlorides, bromides, and iodides of the metals all show a nature analogous to salt. Chloride, bromide, and iodide of silver are particularly interesting to us. These three salts may be obtained by the direct action of chlorine, bromine, and iodine on silver; but a more rapid method is to dissolve in water chloride, bromide, or iodide of sodium, and to add to them a solution of a salt of silver. For silver can also form salts.

If a silver coin is thrown into nitric acid, it is dissolved, forming nitrate of silver; and this is obtained on evaporating the solution as a white soluble salt, which when fused is called lunar caustic.

If a solution of this substance be mixed with a solution of chloride of sodium, a white curdy precipitate of chloride of silver is formed by double decomposition. Chloride of sodium and nitrate of silver produce chloride of silver and nitrate of sodium.

Bromide or iodide of silver may be produced exactly in the same manner if bromide or iodide of sodium be added to a solution of silver.

Bromide, chloride, and iodide of silver are thus separated as precipitates, because they are all three insoluble in water. After being washed and dried, all three salts appear in the form of powders, the chloride being white, the bromide yellowish white, and the iodide yellow. All three are very stable bodies, not decomposed by heat, and insoluble in water, alcohol, or ether; they are, however, dissolved by solutions of hypo-sulphite of soda or cyanide of potassium, by combining with these bodies to form new chemical compounds which are soluble in water.

These three stable compounds—chloride, bromide, and iodide of silver—show a marked sensitiveness to light, and this sensitiveness is the basis of modern photography.

By the light of a gas lamp in a dark room the chloride of silver appears perfectly white, but it quickly takes a violet tint in the daylight. It is often said that it becomes black; this, however, is not the case. This change of colour is the result of a chemical decomposition. The chlorine is in part set free, and may be detected by its odour if large quantities of chloride of silver be employed. The violet powder which remains behind was formerly thought to be metallic silver.

Metallic silver does, it is true, under certain circumstances, present itself in the form of a grey or violet powder. The violet-coloured body formed on exposing chloride of silver to light is not, however, metallic silver; it is only a combination of silver with chlorine, which contains half as much chlorine as chloride of silver. Silver and chlorine form two compounds—one white and rich in chlorine, the other violet and with little chlorine, named subchloride of silver. In the same manner, silver forms two compounds with bromine—one light yellow, rich in bromine, named bromide of silver; and a yellowish-grey compound, less rich in bromine, named subbromide of silver. Further, analogous to these there exist a yellow iodide of silver, and a green subiodide of silver, less rich in iodide. Subbromide and subiodide of silver are produced exactly in the same manner as the subchloride, by the operation of light. The chemist says, therefore, that bromide, chloride, and iodide of silver are reduced by the action of light to the corresponding subchloride, subbromide, and subiodide.

The change of colour by which this chemical change is accompanied is most striking with chloride of silver, less with bromide of silver, and least so with the iodide.

It would appear from this that chloride of silver is the most useful to photography. But this is not the case. We have previously seen, whilst discussing the practical part of photography, that plates of iodide of silver and not of chloride of silver are exposed in the camera. The image thus produced is nearly invisible, but becomes visible through a subsequent process, named the developing process.

In daguerreotypes the exposed plate of iodide of silver was exposed to the vapour of mercury. In this case the vapour was condensed in fine globules on the exposed places, in proportion to the change caused by the light. In the present treatment with collodion, the plate is washed over with a solution of green vitriol. This mixes with the adhering solution of silver, and precipitates from it a fine black silver powder, which adheres to the exposed places of the plate.

Therefore, in both cases we have a finely pulverized body, which is attracted and retained by the exposed places—a mysterious process, as interesting as it is practically important.

From this it appears that it is by no means the colouring of the silver salts which renders the image visible, but the subsequent developing process.

If an experiment be made with chloride, bromide, and iodide of silver simultaneously, by exposing and developing them, it is found that chloride of silver gives the feeblest picture under the developer, bromide of silver a stronger one, and iodide of silver the strongest. Therefore, the very body which was most strongly coloured by light is the least coloured under the developer, and the body which is least coloured by the light, viz., iodide of silver, is the most coloured under the developer.

The developing process is of immense importance. If it were attempted to produce a picture by exposure in the camera without developing, an exposure of hours would be required before the impression could be seen. The

developing process permits, under favourable circumstances, the impression to become visible after an exposure of only one-hundredth of a second.

Pure iodide of silver was formerly used in photography, but a mixture of iodide and bromide of silver is
now preferred. This change was made because it was
soon perceived that iodide of silver is very sensitive to
strong light, but by no means so to weak light. For
example, in taking a portrait, iodide of silver gives the
light parts in a few seconds with great clearness, such
as the shirt and the face; whereas the darker parts, such
as the shadows, the dark coat, etc., are very feebly given.
But, if some bromide of silver is mixed with the iodide,
the coating of combined iodide and bromide of silver
gives a weaker but still intense picture of the clear parts,
while it gives a much better impression of the dark parts
than iodide of silver alone.

The mixture of iodide and bromide of silver is effected
in practice by adding to the collodion a salt containing
iodine and a salt containing bromine; for example, iodide
of potassium and bromide of cadmium. Both are decomposed in the silver bath. Iodide of potassium and
nitrate of silver produce iodide of silver and nitrate of
potash, and in the same way bromide of cadmium and
nitrate of silver produce bromide of silver and nitrate of
cadmium.

A considerable quantity of the solution of silver
remains also adhering mechanically to the collodion
coating. This adhering solution of silver is by no means
a matter of secondary importance; on the contrary, while
developing, it affords the necessary material from which
the fine silver powder is precipitated.

If the developer (for example, a solution of green vitriol) is mixed with a solution of silver, the silver is precipitated in the form of a fine powder. For green vitriol—protosulphate of iron—readily absorbs oxygen, and is changed thereby into persulphate of iron. Accordingly, if a body containing oxygen (for example, nitrate of silver) is mixed with green vitriol, the latter withdraws at once the oxygen from the silver salt, and sets free the silver. Other bodies that readily combine with oxygen operate in like manner; namely, certain organic substances, such as pyrogallic acid and others. It was formerly thought that green vitriol reduced the iodide of silver affected by light; and this erroneous opinion is actually found in some of the most recent works on chemistry. It can be easily proved that this view is false. For, if a plate is exposed and the nitrate of silver adhering to it is washed away, and then the developer poured upon it, no picture appears, proving that green vitriol alone is without action on exposed iodide of silver. But if a solution of silver is added, a picture appears immediately.

The solution of silver adhering to the plate plays, however, another part. If a plate is washed before it is exposed—that is, if all the nitrate of silver which adheres to it is removed, and it is then exposed—it will be remarked that it is far less sensitive than when the nitrate of silver is present.

This is explained by the peculiar property of many bodies sensitive to light.

There are bodies which in isolation are either not at all, or only very slightly, sensitive to light, but which become so in the presence of substances which are able to unite with one of the constituents liberated during exposure to light. For example, chloride of iron is not sensitive to light; but chloride of iron dissolved in ether is sensitive, because the liberated chlorine unites at once chemically with the ether.

The same remark applies to iodide of silver. This is, by itself alone, sensitive to light, but only slightly; in presence of a body which can combine with iodine, it is quickly decomposed in the light. Now, nitrate of silver, which reacts with iodine with the greatest ease, satisfies this condition; and this explains the greater sensitiveness of iodide of silver in the presence of nitrate of silver.

It follows from this fact, which was first accurately determined by the author, that other bodies which unite easily with iodine also increase the sensitiveness of iodide of silver.

Among these bodies may be enumerated extract of coffee or tea, morphine, and tannin. Such bodies enable photographers to prepare what are called dry plates. The plates, which are prepared in a silver bath, only remain moist for a short time; the adhering solution of silver dries up, and then dissolves the iodide of silver, so that the plate is actually eaten into. It is not therefore possible to keep a supply of such plates for any length of time, which would be very useful in travelling.

But dry plates which may be kept without spoiling can be prepared by washing away the nitrate of silver adhering to the moist plate, and then coating the plate with a solution of a substance having affinity for iodine; for example, with tannin or morphine. Such coatings can dry up without injury to the film of iodide of silver, and in this manner a durable dry plate is obtained. The sensitiveness of the plates is considerably less than that of moist plates, but this is of no detriment in the case of objects well illuminated. The development of dry plates of this kind is commonly effected with pyrogallic acid. This substance is obtained by dry distillation of the gall-nut. It has a powerfully reducing action ; that is, it precipitates metallic silver from silver solutions, exactly as green vitriol does.

But pyrogallic acid alone is not able to bring out an image on an exposed dry plate, because another substance is necessary to yield the silver. This substance, viz., a solution of silver, is found on the plates themselves when they are wet. But in the case of dry plates, the silver salt has been washed off; therefore mixed solutions of pyrogallic acid and nitrate of silver must be employed as developer. Finely divided silver is precipitated, adheres to the exposed places, and thus brings out the image. Nevertheless, dry plates do not give such beautiful and secure results as moist plates.

We have now given an illustration of the photo-chemical phenomena in the production of a camera picture. The essential part of this process—the negative process—consists in the developing of an invisible impression made by light by a subsequent operation.

But all pictures are by no means prepared in this way. We have already seen, on the contrary, that the pictures on paper are occasioned by the production of a visible impression of light, a piece of sensitized paper being exposed until it is coloured dark. In this case no developing is required. The picture is exposed to the light till it has received the necessary intensity.

The process is in this case quite simple. The paper

I

contains chloride of silver and nitrate of silver. The former is quickly, the latter slowly, reduced by the light to metallic silver, which is separated as a brown powder. Chloride of silver alone would only be reduced to sub-chloride of silver; but in the presence of paper-fibre the process of reduction is carried further, and metallic silver is formed. The chlorine set free by the light combines immediately with the silver of the nitrate of silver, and produces chloride of silver again. This is then decomposed by the light, and a fresh quantity of brown metallic silver is separated. And this process is repeated as long as nitrate of silver is present, and as long as the light operates.

Pure chloride of silver alone gives but a faint impression, but in presence of nitrate of silver it yields a very vivid image. The picture, in the form in which it is produced by the light, is not durable—it would turn brown through the further operation of light on the white places; and to prevent this, the sensitive salts of silver still adhering to the paper must be removed. The nitrate of silver is removed easily by washing with water, for it is soluble in water; but the chloride of silver must be removed by plunging in a solution of hypo-sulphite of soda. This salt reacts with chloride of silver, forming chloride of sodium and hypo-sulphite of silver, and the latter combines with the hypo-sulphite of soda to form a double salt, remarkable for its peculiar sweet taste, which is soluble in water, and can be removed by washing.

If a fresh print is plunged in hypo-sulphite of soda, it suddenly changes its beautiful violet colour—it becomes a yellowish brown; this tint is not liked.

It does not interfere with the effect in technical and scientific pictures, but is a great drawback in portraits and landscapes; so the positive prints are subjected to a further treatment, styled the toning process. They are immersed in a very dilute solution of chloride of gold. Metallic silver has more affinity for chlorine than has gold; hence it combines with the chlorine, forming chloride of silver, while the gold is precipitated, giving a blue colour to the picture. And this blue, mixed with the brown of the picture, gives a pleasant tone, which does not change in the fixing-bath; that is, in hyposulphite of soda.

Accordingly, every paper photograph consists of silver and gold, in the proportion of about four parts of silver to one of gold; the quantity of both substances being very small. In a picture of 44×47 centimetres, or 17 inches by 22, only one-thirteenth of a gramme, or about one grain, of metallic silver is contained. Its value is about one German pfennig,* and the value of the silver in a carte de visite is about one-thirtieth of a pfennig. The question may here arise, how it happens that photographers charge so high for their pictures? A sufficient reply is found in the fact that the price is not determined by the value of the materials, but by the labour which has been necessary to produce the pictures. It must be remembered that a photographer has to make twenty-eight operations to produce a negative, and eight to produce a positive; that a picture is often a failure; that for the one pfennig's-worth of silver in preparing a sheet, four-pennyworth of salts of silver must be employed, and that at the utmost only one-third of this

* About half an English farthing.

silver can be recovered from the washings. Nor should
it be forgotten that the paper itself is valued at three-
pence, that cardboard of the same value is required for
the mounting, and that further outlay is needed for the
hire of premises and assistants—all which circumstances
certainly justify the price demanded.

If it is borne in mind that thirty-three times as much
silver must be employed as that which actually remains
when the picture is finished, it will be seen that the
amount of silver consumed annually in photography must
be enormous.   It is valued at about £1,350,000.

CHAPTER XII.

ON THE CORRECTNESS OF PHOTOGRAPHS.

Influence of the Individuality of the Photographer—Different Branches
of Photography—Influence of Lenses, of the Length of Exposure,
of Colours and Models—The Characteristic Feature in the Picture
—Deviation from Truth in Photography—Difference between Photography and Art.

(a) *Influence of the Individuality of the Photographer.*

IN the previous chapters we have become acquainted
with the development, theory, and practice of photography. We have mentioned cursorily various practical
applications; for example, the *lichtpaus* process. It is
our present purpose to give our special attention to one
point which is of great import in judging of the value
of a photograph.

Most persons have a fancy that the application of
photography is always uniform, whatever may be the
object to be taken, and, therefore, that a photographer
who can take a portrait must be able to take equally
well a machine, a landscape, or an oil-painting. This
results from the erroneous notion that the picture makes
itself when the photographer removes and replaces the
cap of the lens. But our readers know already that the
picture does not make itself, but that it must be first

developed, intensified, fixed, and printed. In all these operations there is no precise measure or rule how long the photographer should expose to the light, develop, intensify, print, or tone his picture; it rests entirely upon his discretion. He is able at pleasure to bring out more or less of the details of a picture by abridging or prolonging the exposure, to make a picture more or less brilliant by intensifying, more or less dark by regulating the printing and toning. He must direct his judgment as to the correctness of his picture by nature and by nature alone. He must know nature, and compare his picture with it. Nor is this easy. Nature appears positive to him, but in the picture she at first appears negative; and if he compares the two, he must be able to convert the picture in his mind into a positive, which it is afterwards to become. More experience and study are required to do this than is generally supposed.

If two printed proofs are presented to a man who is ignorant of the art of printing, one of the sheets in question being well and the other ill printed, if the defects be not too glaring, this person will not be able to detect any difference between the proofs. The practised eye of the printer, however, immediately detects that in one proof the type is too thick or thin, or that the letters are faint or uneven. In like manner, a practised eye is needed to judge a photograph—an eye not only able to observe the finest details of the picture, but also the peculiarities of the original. The unprofessional man often uses the expression, "I have no eye for it,"—that is, "I am not accustomed to see such things,"—and it is in this manner that we first discover how imperfectly we use this, the most perfect of our senses.

A man born blind, and who recovers his sight by an operation, cannot at first distinguish a cube from a sphere, or a cat from a dog. He is not accustomed to see such things, and must first learn to see.

We, also, though in possession of sound eyes, are blind to all things that we are not accustomed to see; and this fact is most apparent in art, as also in photography, so closely related to it.

If photographers principally engaged in taking portraits are not able to produce a good landscape, the reason of this is that they have no eye for landscapes; that they consider a picture to be good after too short an exposure, or when imperfectly developed and intensified, or when badly printed. It proceeds from their not knowing the influence exercised by the position and intensity of the sun or clouds, without speaking of other points of less importance.

Thus every class of subjects requires a special study, although the mere manipulation remains in all cases the same. There are photographers whose proper province is portraiture, and others devoted to landscapes, to the reproduction of oil-paintings, etc.

(b) *Influence of the Object, of the Apparatus, and of the Process.*

The remark is frequently made by admirers of photography, that this newly invented art gives a perfectly truthful representation of objects, understanding by the term truthful a perfect agreement with reality. Photography can, in fact, when properly applied, produce truer pictures than any other art; but it is not absolutely true. And, as it is not so, it is important to become acquainted

with some of the many sources of inaccuracy. I shall
treat here especially of optical errors.

The lenses which are employed in photography do
not always give absolutely true pictures. Suppose, for
example, that an image of a square is formed by a simple
lens ; the sides of the image will often be curved, as in the
accompanying diagrams, though not so markedly. A picture taken with such a lens, in which straight lines come
out as curves, is evidently inaccurate. The inaccuracy
may not be perceived by many, but it exists. It may per

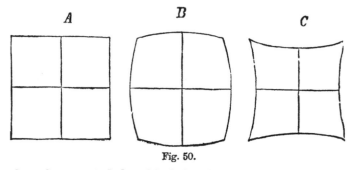

Fig. 50.

haps be expected that this defect disappears in the case of
what are called correct lenses, but let the attempt be made
to obtain a picture with these correct lenses of lofty buildings taken from a low position. The lines that ought to
be perpendicular commonly converge upwards. This is
caused by the photographer being obliged to direct his
camera upwards, in order to be able to take in a view of
the whole building, and thus the perpendicular lines are
represented as convergent. To avoid this defect, lenses
have been made with a very large field of view. These
are called pantoscopic ; but they reproduce distant objects

apparently on a very small scale, and objects near at hand on a very large scale—peculiarities unnoticed by unprofessional persons, but detected by close observers of nature.

A remarkable phenomenon, exciting the wonder of the uninitiated, is the distortion of spheres in photography. Let the reader imagine a row of cannon balls: these will always appear spherical to the eye, and an artist would always draw them as spheres; but if they are taken

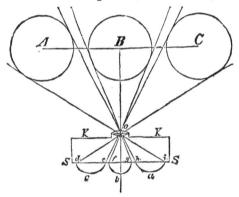

Fig. 51.

with a lens with a large field of view, the outside balls no longer appear circular but elliptical.

To explain this phenomenon, we must attend once more to the mode in which the picture is produced. Let it be conceived that there are three balls A B C in front of a camera K, with the lens o (Fig. 51). Each ball projects a cone of rays on the optical centre of the lens, and an inverted cone is formed within the camera, and cuts the surface of the picture, if its axis falls obliquely upon it, in the form of an ellipse, such as A and C. It is only when the axis of the cone of rays is perpendicular to the surface of the picture $S\,S$, as at B, that the image appears a circle. This defect, however, only occurs when the field of view of the lens is very large, and the balls are situated very near its border.

A photographer once brought to the author the picture of a castle having a row of statues in front of it, which he had taken with a lens having a large field. Curiously enough, the heads and bodies of the statues towards the margin of the picture became broader and broader; and the slim Apollo of Belvedere, which unfortunately stood on the very margin, had such full-blown cheeks and so protuberant a paunch, that he looked like Dr. Luther.

But, quite independently of these considerations, there is another point that must materially affect the accuracy of photographic representation. Photography generally gives the light parts too light, and exaggerates the dark shadows. This is a natural fault, which it is very difficult to avoid. It is seen in the most apparent manner in taking objects lighted by a brilliant sunlight; for example, a statue. If the exposure is short, a detailed picture is obtained of the light side, but the shady side is a black blotch. If the exposure is long, the shady side is full of detail, but the light side exposed too much, and so deeply coloured that the details are wanting in it. Hence photographers, in order to obtain correct pictures, are obliged to avoid strong contrasts of light and shade in their models far more than would be thought desirable by artists. The latter often exclaim when they see the illumination of a model, and are surprised when the picture is correct. And, no doubt, in the case of landscapes and architecture, the results are not always satisfactory.

The author once took a photograph of the interior of a laboratory which had a vaulted roof. All was excellent. The tables, furnaces, retorts, lamps, etc., were all seen, only the vaulted ceiling was quite dark. New attempts were made, with exposures of 20, 30, and 40 minutes. At length a trace of the vault appeared; but now the objects near the window were suffering from too much exposure; that is, they had become as white as if they had been snowed over. This circumstance, that photography exaggerates the dark parts, appears again in very simple matters, such as the reproduction of copper-plate engravings. A photographer once reproduced a painting of Kaulbach's "Battle of the Huns." He produced a charming photograph, but the city in the background appeared too thick and black, and not soft enough. The customer refused the photograph and demanded another. The photographer made another attempt, giving a longer exposure, and now the distance appeared softened down; but, unfortunately, the objects close at hand, which had to appear black and clear, turned out grey. In the end, the photographer escaped from the difficulty by retouching the negative. These are quite ordinary examples to show how difficult it is to reproduce an object correctly.

But we come now to the worst point, the colours. Photography gives the cold colours—blue, violet, and green—too light, and the warm colours too dark. Take as an instance the photographs on sale of "A Sunset on the Ganges," by Hildebrandt. It represents a red glowing sun with clouds of chrome yellow on an ultramarine

sky. But what becomes of all this in the photograph ?
A black round disc between black thunderclouds. It
looks like an eclipse at Aden. The difficulty of representing nature is still more patent when the photographer
attempts to grapple with higher artistic themes. Let us
take an example. There exists a pretty *genre* picture
called " A Mother's Love." A young mother sits reading
in an armchair; her little darling embraces her suddenly
from behind, and, delightfully surprised, she drops her
hand with the book, turns to look at her little pet, and
offers him her cheek to kiss.

A photographer was inspired with the idea of producing a similar picture with the help of a living model,
He found a comely maiden, who agreed to personate the
mother, and a good-looking boy was also found. An
armchair for the mother, a chair, and other suitable
furniture were easily procured. The next point was the
grouping. The pseudo-mother was very accommodating
to the requirements of the photographer, and even assumed
a look which, for want of a better, might pass as the
expression of a mother's love. But the boy was not of
the same mind. He was by no means attracted by the
pseudo-mother—he protested against coming near her,
and a good cuff was needed to make him take up the
requisite position. Time was thus lost. The mother
began to feel uncomfortable in the irksome position,
straining her neck. The photograph was taken at last,
and turned out sharp and without spot or blemish. The
models were dismissed to their great satisfaction. What
was the result ? The boy was embracing his mother with
a face bearing evidence of the cuff he had received, and
with a look as if he would have liked to murder her;

and she regards him with an expression that seems to say, "Charles, you are very unmannerly," and appears greatly annoyed that her pleasant reading has been interrupted. Can it be said that a picture of this kind correctly expresses the intention of the artist ?   Does the picture thus produced correspond accurately to its name, " A Mother's Love " ?   The untruthfulness of such a picture will be evident to every one.

Thousands of pictures of this class are offered for sale. About ten years ago errors of this kind were committed by the thousand in stereoscopic views, and if they meet with approval this must be referred exclusively to the bad taste of the public.   It may, however, be said in this case it is not the photographer who is guilty, but the unwilling models.

Nevertheless, it is this very circumstance that throws such immense difficulties in the way of taking good photographic portraits.   Many persons by no means wish that their characters should be correctly given.   The rascal wishes to appear an honourable man in his picture ; tottering old men desire to seem young, foppish, and lively ; the maid-servant plays the fine lady, the trades-man's daughter would be a court lady, the crossing-sweeper a gentleman.   Thus the picture serves them only as a means of flattering their personal vanity ; and, in order to appear very noble and distinguished, they put on a Sunday dress, often borrowed and a very bad fit. They practise at home, moreover, before their looking-glass, in the presence of papa, mamma, wife, or lover, atti-tudes impossible in an artistic point of view.   Even culti-vated persons are not exempt from these absurdities. Thorwaldsen relates of Byron, who gave him a *séance* :

"He sat down opposite to me, but assumed, immediately I commenced, a perfectly different expression. I called his attention to this. 'That is the true expression of my face,' replied Byron. 'Indeed,' I rejoined, and then made his portrait exactly as I wished. All persons declared my bust to be an excellent likeness. But Lord Byron exclaimed, 'The bust does not resemble me; I look much more unhappy.' The fact was that at that time he wished to look intensely miserable," adds Thorwaldsen. The photographer is even in a worse case. If Byron had come to a photographer and had presented his face of misery to the camera, what could the photographer have done? He is unfortunately dependent on the model, and many models leave him in the lurch at the critical moment, often not intentionally, but from nervousness or inadvertence. Much depends here on the influence of the photographer, who must know how to control his sitters with courtesy; many portraits, however, fail without any fault on his part. The author has often witnessed how persons of his acquaintance, at the moment of being taken, assume quite a strange expression without being in the least aware of it.

There are still more characteristic cases of photographic inaccuracy which cannot be attributed to the models. Let us suppose that a photographer, stimulated by the beautiful pictures of Claude, Schirmer, and Hildebrandt, wished to photograph a sunset. He evidently can only expose his plate for a moment to the dazzling bright sun. What sort of picture is the result? A round white blotch and some shining clouds are all that appear clearly. All objects in the landscape—trees, houses, and men—have had too short an exposure, and

form a black mass. There, where the eye clearly distin-
guishes road, village, forest, and meadow, it sees in the
photograph nothing but a dark patch without any outline. Is such a picture true? Even the most fanatical
enthusiast of photography will not dare maintain this.

Cases where violent contrasts of light and shade make
the production of a correct picture quite impossible, are
countless in number. Let any one examine the majority
of the photographs of the white Königsdenkmaal in
the Thiergarten at Berlin. The monument is excellently given, but the background of trees is a confused
black mass, without details, without shades of tone; the
architecture and other features are there, all except the
splendid foliage that delights the eye at that spot. Still
more numerous are the photographs of rooms, in which
the dark corners, quite discernible to the eye, present
nothing but pitch-black night.

There are other cases besides these of photographic
incorrectness. Suppose we are looking at a mountain
landscape. A small village, enclosed on both sides by
woody hills, occupies the centre, its houses extending
along the declivities and scattered picturesquely among
the trees. A ridge of finely broken mountains in the
background, their summits shining in the setting sun,
completes the wonderful picture, whose effect is only
injured by one object—a ruinous pigsty close to the
spectator, with a dung-heap beside it. An artist, who
wished to paint this scene, would certainly have no
scruple about altogether leaving out the pigsty, or
leaving it so indistinct and dark that it would not injure
the landscape. But what is the photographer to do?
He cannot pull down the offending object. He seeks

another position; but there the greater part of the land-scape is concealed by trees. He ends by admitting the pigsty, and what kind of picture is the result? On account of its vicinity, the pigsty appears of colossal size in the picture. On the other hand, the landscape, which is the principal thing, appears small and inconsiderable. A still more fatal adjunct is found in the dung-heap occupying almost one-fourth of the picture. As the most brightly lighted part of the photograph, it imme-diately attracts the eye of the beholder; it diverts his glance from other important points. The photograph obtained does not appear to be a picture of the landscape, as intended, but a view of the pigsty. The accessory has become the principal point. The picture is untrue. It is untrue, not because the objects it represents were not present in nature, but because the accessories are pre-sented too glaringly and too large, while the principal features appear too small, indistinct, and inconsiderable.

This brings us to a weak point in photography. It represents accessories and principal features as equally defined. The plate is indifferent to everything, while the genuine artist, in reproducing a view of nature, gives prominence to what is characteristic, and entirely keeps under or softens off accessories. He can dispose and manage his picture with artistic freedom, and he has a perfect right to do so, because, by his giving prominence to what is characteristic, and dropping what is accessory, he is truer than photography, which gives equal promi-nence to both, and often more to what is accessory. Reynolds says of the portrait of a lady in which an apple-tree was most carefully painted on the background: "That is the picture of an apple-tree and not of a lady."

Similar remarks might be made on seeing many photographs. It is a cardinal error in their case, that they
give a stronger tone to accessories than to essentials.
They present a conglomerate of furniture, and it is only
after careful inspection that a man is detected among it
whose portrait should form the picture.   In another case
a quilted white blouse is seen, and it is only after some
time that a girl's head is perceived rising above it. A
park is seen in a landscape, with fountains and other
adornments, and it is only after some time that a black
coat is seen confounded with an equally dark bush.

It may perhaps excite surprise that the writer ascribes
greater truth to painting than to photography, which is
generally regarded as the truest of all methods of producing pictures. It must be self-evident that this remark
can be made only of the works of masters.   One of the
great services of photography is that it has rendered impossible those daubs of inferior artists formerly offered for
sale in every street. But the perfect picture of the photographer is not self-created. He must test, weigh, consider, and remove the difficulties which oppose the production of a true picture. If his picture is to be true, he
must take care that the characteristic is made prominent
and the accessories subordinate.   The photographer attains this end, on the one hand, by appropriate management of the original; on the other, by a proper treatment
of the negative. To do this he must of course be able
to detect what is characteristic and what accessory in his
original.   The sensitive plate of iodide of silver cannot do
this. It receives the impression of all that it has before
it, according to unchangeable laws.

Therefore, whoever wishes to be successful in photo-

K

graphy must first become familiar with the object that he wishes to take, that he may know what he has to do. The photographer will not, indeed, be able to control his matter like the painter, for the disinclination of models and optical and chemical difficulties often frustrate his best endeavours; and hence there must always be a difference between photography and a work of art. This difference may be briefly summed up by saying that photography gives a more faithful picture of the form, and art a more faithful picture of the character.

CHAPTER XIII.

LIGHT, SHADE, AND PERSPECTIVE.

The Difference between the Picture and Reality—Effect of Shade—
Perspective Foreshortenings—Effect of the Position of the Spectator—Influence of Distance—Influence of the Height of the Eye.

IN the previous chapter, while treating of the incorrectness of photographs, we tacitly assumed that it is possible

Fig. 52.

to give a true picture of an
object, if not by photography,
yet by the hand of a skilful
artist.

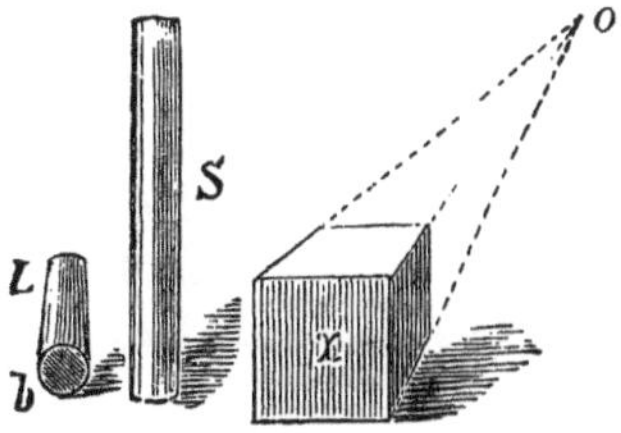

We will now see how far
this assumption is admissible.

Let the simplest case be
taken; for example, a cube
or a cylinder. Let these be drawn, and figures will be
obtained nearly identical with those marked X and S in
the diagram. Now, these figures are flat like the paper,
while the originals are solids. It may be said that
picture and solid agree; but it is not so. Let a blind
man be questioned, who knows the bodies by touch
only. Now, the cube can be moulded in marble or plaster,
and then the deception—for such it is—can be carried to

great lengths. The wood of the cube or of the cylinder can be imitated by painting. The eye will readily pronounce such imitation to be wood. The blind man, who feels both, will say : The form agrees, but not the mass — one cube, that of wood, feels warm ; the other, that of stone, cold.

The principles that apply to these two objects apply to all objects and their representations. No one of them is a perfectly true copy of the object. When the surface representation makes on our eye the impression of a solid object, this is an illusion by which our eye suffers itself to be deceived.

If two rectangles A and B are drawn on paper, both appear as plane figures. But directly one of them B is shaded with thinner or thicker lines, the rectangle no longer appears flat, but cylindrical. Thus, by imitating

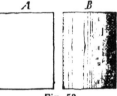

Fig. 53.

the gradation of light and shade, we have produced a deception for our eye. This division of light and shade is one of the most important means of producing an appearance of solidity.

But there is another and a more important means of deception—perspective.

If we observe the cube (Fig. 52), the faces of which are equal, we perceive that these faces appear of very different length. The surface turned towards our eye appears a square, while the others are shortened in a marked degree, the surface appearing quite irregular, the parallel lines running together and converging to one point o, called the vanishing point. The same thing happens with all other bodies: a human arm hanging down or a standing column S (Fig. 52) appear at their full length, but the lying column L, and the arm extended towards us, appear foreshortened. Their dimensions are contracted; in short, we see, instead of the shaft of a column, only its circular base b, and this, again, appears sometimes round, when its full surface is turned towards us, at others an ellipse, which it is not in fact, and in this case the parallel sides of the column converge. The track of a railroad viewed in perspective presents the same features. The fact that we do not feel this deception—for such it is—to be one, results from habit.

We know from experience that the arm extended towards us is longer than it appears, in perspective, to our eye, and also that the rails which appear to run together are parallel. We are continually correcting the errors of our sense of sight. The eye gives us a false representation of objects, and the painter takes advantage of this circumstance. He represents the lying column $L\ b$, and the sides of the cube, as falsely as we see them—that is, "foreshortened" in their dimensions, with their parallel lines converging—and every one is deceived by this.

It is the task of the artist as of the photographer to represent perspective correctly; that is, as it appears to our eye. If this is not the case, the picture appears incorrect.

Perspective teaches us the laws of foreshortening.

Our eye is a camera obscura with a simple landscape lens. It is known from optics, that the image of any point lies on the straight line drawn from the point to the optical centre of the objective, at the place where this

line, named the principal radius, cuts the plane of the
image—the ground-glass screen of the camera or the
retina of our eye. The image of a straight line is the
place where the rays from each point of the line, passing
through the optical centre, cut the ground-glass screen.
Now, these rays form a plane, and this plane cuts the flat
screen in a straight line. Therefore the image of a
straight line is to our eyes another straight line, and the
image of a plane triangle another plane triangle. If
the flat figure is parallel to the retina, by well-known
stereometric laws the image is like the original. Let the
reader imagine a glass slab placed perpendicular to

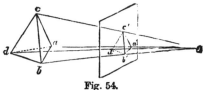

Fig. 54.

the axis of his eye; then the rays or pencils of light
issuing from this object a b c d will cut it so as to form a
figure a' b' c' d' (Fig. 54). If such a figure is drawn for
a given point of intersection, this drawing, if brought to
a proper position and distance from the eye, will produce
on it exactly the same impression as the object itself.
This is the secret of the solid appearance of plane pictures
properly constructed. A picture made in the manner
just described is named a drawing in perspective. It is
evident that such a drawing must be viewed under the
same conditions as those in which it was designed.

If *A B C D* (Fig. 55) is the outline of a house, *B* the
picture, *O* the point of intersection of the rays, and *a b c d*

the image of the points *A B C D*, the eye must be brought exactly to the point of intersection *O* if the representation in perspective *a b c d* is to produce the same impression as the object.

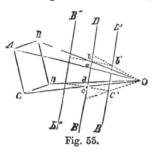

Fig. 55.

If the picture is brought nearer to the eye (for example, to *B'*), it is evident that the rays will intersect at a very different angle from those issuing from the object *A B C D*; accordingly, they cannot produce a correct impression. The same thing would be the case if the picture were removed farther from the eye (*e.g.* to *B''*). Therefore every drawing in perspective must be viewed from the point of intersection of the rays adopted as the basis of its construction, if it is to produce a correct impression.

Now, photography is a drawing in perspective whose point of sight is in the objective. Accordingly the inspecting eye must be brought to the same distance as the objective; that is, to the focal distance. If this is not done, the impression is untrue.

We have lenses with a focal distance of only four inches, and even less; and at such a distance it is impossible to see a drawing with the unaided eye. To do this it must be held at the distance of at least eight inches, and that is the reason why photography in such cases produces an untrue impression. This is often the case with views taken with lenses of great aperture.

There are other abnormal appearances which accompany portrait taking. Thus, the same object gives an entirely different picture according as it is viewed from a greater or less distance. Let the reader conceive a pillar with the outline *A B C D*, let it be viewed from *P;* in this case the faces *A B* and *C D* will be perfectly seen. Now let the spectator approach nearer to the object; for example, to *O.* From this position nothing is any longer seen of the faces; the entire character of the picture becomes changed. If instead of a pillar a human face

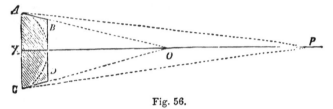

Fig. 56.

be thought of, it is evident that the cheeks will contract if we approach the object, and the face appear too narrow in proportion to its height.

The accuracy of this conclusion is proved by the following illustrations. The two representations (Fig. 57) of the head of Apollo were taken at the distance of 47 and 112 inches.* The bust was placed perfectly upright, also the photographic apparatus, and the directing line was most carefully arranged.

The contrast is obvious. The whole figure appears in I. slimmer, the chest almost contracted; on the other hand, the same model II. appears with full cheeks and square shoulders. That this slimness is by no means a

---

* In order to secure a correct reproduction of both, they were reproduced upon wood by a process of photographic wood engraving. The reproduction does not give the same impression as the original, but it is sufficiently intelligible to the careful observer.

mere deception of the eye may ascertained by measurement.*

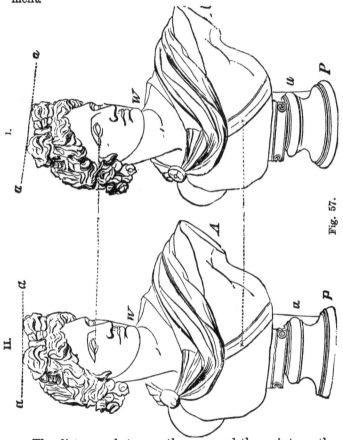

Fig. 57.

The distances between the eye and the point on the

* In the original photograph, where the two busts stand out from a black background, this difference is still more marked.

chest marked by a cross are exactly equal in both busts
—the greatest breadth of chest, including the arms,
amounts in I. to 56 millimetres, and in II. to 59 millimetres. Quite independently of this glaring difference,
there are other marked distinctions between the two
heads which strike a careful observer. Let a line a a be
drawn across the top of the hair—in II. it is horizontal,
in I. it inclines to the right.

Next let attention be directed to the pedestal. The
curves in I. are strongly inclined, and in II. are quite
horizontal.

Let the ends of the arms A A be next considered.
In I. the side surface is scarcely seen, and in II. it is very
apparent. In like manner, it is clearly seen that the
back pedestal at u stands out more in II. than in I. In
II. the head stands more between the shoulders (let the
angle of the neck be observed at W); in I. it rises up
more; therefore the whole form appears in I. to raise up
the head. In II. the head appears somewhat bent forward; and yet the figure was immovable, the lenses
employed free from flaw, the direction and height were
the same in both. Nothing was different but the distance.

The author, besides taking these two heads, has taken
two others at the distance of 60 and 80 inches; and if
the four heads thus taken are placed beside each other, it
is seen how with the increase of distance the form becomes fuller and more thickset; how the hair sinks more
and more; how the ellipses of the pedestal become flatter;
how the chest increases in width, and the stumps of
the arms stand out.

Thus, therefore, we see very different views of the

same object at different distances; just as the same portrait placed in different lights expresses an entirely different character.

It may be objected that these are small matters, and that it is indifferent whether the statue looks a little too thin or too stout. To many this may appear unimportant in the case of Apollo—most persons do not know in the least how he looks. But it is a different matter in the photography of portraits, when the personality of the customer himself is in question. Persons quite untutored in art have a very quick eye for their own physiognomy—a line, a wrinkle, an outline, a curl, are in this case criticized, and differences that would not be at all remarked in the picture of Apollo become very striking It is therefore the duty of the photographer to attend to the effects of distance.

Now, many persons would perhaps wish to know which distance is the best; which gives the most correct picture.

We might reply that this depends on the individuality of the person. Painters in general recommend for the drawing of an object a distance that is twice its own length; accordingly, for a man six feet in height, a distance of about twelve feet; for his bust, about five feet. The painter, however, has here greater freedom; he can add, omit, and alter at his pleasure. In photography this is only partially possible.

The appearance of hollow bodies is as much altered by distance as that of solids.

If $A\ B\ C\ D$ (Fig. 58) is the inside of a box, we should see the side $A\ B$ much more foreshortened from the distance P than from O' or N; therefore, its picture taken at short distances will show it wider in proportion to its height than if taken from a more distant point. The

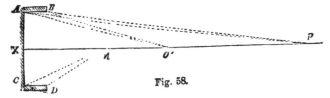

Fig. 58.

same thing occurs if we imagine $A\ C$ to be the trunk, and $C\ D$ the lap or the feet of a seated person. In that case the lap appears much larger in relation to the trunk, and the feet of a standing person appear longer from the shorter distance. Let the reader observe, for instance, the foot

Fig. 59.   Fig. 60.

of the Apollo in I. (Fig. 57), which is much more prominent than in II. Lastly, let $C\ D$ (Fig 58), be supposed to be the carpet or ground; this will appear wider—that is,

rising higher—seen from N.   Therefore, if the same
person is taken from different positions, P and O', so that
the height of the body remains the same in both pictures,
in that one taken at the shorter distance the prominent
parts—lap, hands, and feet—appear wider, and the ground
or chair more inclined (Fig 59) than in the picture taken
from P.

Fig. 61.

Very important changes result from an alteration in
the height of the spectator's eye.

If a standing person is looked up to, so that the head of the spectator is lower than the head of the object, the latter appears thrown back. If the head of the spectator is on a level with the head of the object, the latter appears perpendicular; if the spectator is higher, the head of the object appears inclined forward.

The three accompanying diagrams, taken from photo-

Fig. 62.

graphs, will make this evident. The first shows the view taken on the same level, the second taken from above, the third from below.

Similar differences occur in viewing a landscape from

a high and from a low position, as may be seen from
the three woodcuts on next page.   The dotted horizontal line shows the height of the eyes of the spectator
(his horizon).   The first picture gives a view as a person
sitting on the ground would see it; the milestone on the
left appears unusually high, towering to the sky, and
the men appear taller, but the ground looks contracted

Fig. 63.

(foreshortened).   The second picture gives the view as
seen by a man standing erect: in this case the ground
widens out, rising higher, and the milestone appears

Fig. 64.

Fig. 65.

Fig. 66.

lower. In the third picture, which gives the view from twice the height of the man, the figures and the milestone appear small and contracted. They appear persons who are smaller than the spectator, while the ground widens out and rises considerably in the picture. These examples show how important is the choice of position both in photography and painting, and how an incorrect choice produces quite an abnormal picture. The photographer is unfortunately often obliged to take a position that does not give a favourable view, for example, of lofty buildings in narrow streets (the Rathhaus in Berlin, St. Stephen's at Vienna), or among mountains, where often the trunk of a tree, which does not offend the eye of the spectator, destroys the picture of the photographer, and forces upon him the choice of a less picturesque but unincumbered locality.

CHAPTER XIV.

THE APPLICATIONS OF PHOTOGRAPHY.

WE have by this time learnt the difficulties of obtaining a correct photographic picture; and the reader is now capable of examining the various problems attempted by photography.

We shall carry this examination only so far as is necesssary for the comprehension of the subject, and is of general interest to every one.

SECTION I.—PORTRAIT PHOTOGRAPHY.

Popularity of Portraiture—Æsthetical Defects—Dependence of Success on the Person to be taken—Effect of Dress—Effect of Colours—Pictures of Children and Groups—Effect of the Size of the Picture—Life-size Pictures—Instantaneous Pictures—Photographic Copies of Photographs.

Scarcely any other branch of photography has enjoyed so much popularity as that of portrait-taking. Most persons understand by the term photography only the art of portraiture, and few only are aware that it can be used for any other purpose. The photographic portrait owes its popularity to its extraordinary cheapness, to the rapidity of execution, and to its relatively greater resemblance when compared with drawings from nature. Photography, with all its imperfections, can reckon, on account of these circumstances, on greater support than the drawing of an artist; and the more so as a false belief exists that photography is invariably accurate, which is by no means the case, as we have already seen.

Photographic portraiture having driven the trade of portrait-painting out of the field, the genuine artist only is able to hold his own against it. This branch—portraiture—more than any other depends on the good taste of the operator—on his power of giving a natural or at least apparently natural position to the sitter, concealing as far as possible all personal imperfections. He must by clever manipulation of the light be able to give prominence to the essential points and leave indistinct those parts which would injure the effect of the picture. To this end the photographer is able to include the surroundings of the person, be they a chamber or a landscape, or to exclude them by screens.

In the first period of photography the pictures were commonly crammed with accessories, and incredible errors were committed with relation to position and light; but now the more advanced photographers have taken hints from artists, and latterly pictures are seen which, notwithstanding the mechanical character inherent in their production, create quite the impression of art.

The model—that is, the person to be taken—has a very essential share in the work of photography. Not unfrequently persons go to a photographer in a very morose frame of mind, or they lose their patience at some delay or other; it also often happens that people go who are slightly unwell—with headache or a bad night's rest. This is a great mistake; the bodily or mental condition stamps itself infallibly on the picture, and often gives it an expression unlike the original, even after the photographer has used all his art upon it. In like manner it very frequently happens that persons at the moment of being photographed put on a perfectly strange expression, force a smile, stare, let the mouth fall open, or are disturbed by the iron rests which keep the head steady, and are quite essential if a well-defined picture is desired, but which are only submitted to under protest by the model, who fancies he can sit motionless without such aid.

None of these influences can be set aside by the photographer. The persons who present themselves to have their portraits taken are for the most part unknown to him. He has often only five minutes to study their faces, to find their best side, and to place them in accord with the surroundings. He probably attends to these matters as skilfully as any one, yet he has no power over the features of the original. Nor has he any idea whether the expression of a person is his usual one, or happens to be modified by ill temper or ill health. In the latter case it is impossible for the picture to please, however masterly may be the execution; yet the fault lies not with the photographer but with the original.

Another cause of failure is that many persons insist upon choosing their own position, whether of their own accord or prompted by friends. All attempts to make a good picture under such circumstances will generally fail, because the errors in perspective previously enumerated are overlooked. People in general do not know this, but the photographer does. Other inconveniences result from the very nature of photography. Blue eyes are generally too light and dim, blonde or auburn hair too dark. Many of these difficulties are removed by clever retouching, but by no means all.

Still greater are the hindrances which the toilette and the caprices of fashion furnish. Bad taste in dress is more visible in photography than in nature. Ladies deficient in taste frequently make their necks, which are naturally short, appear still shorter and thicker by neck-laces. They disfigure a form, perhaps naturally excellent, by long trains; the back of their head by an ill-chosen chignon; and their hair by ribbons of the brightest and most glaring colours. In these cases the photographer can do much service by his good advice. The difficulties are even increased if groups or children are in question.

Children must be amused: and if the photographer wishes to succeed in taking children's portraits, he must know how to win their confidence; this is the reason why many photographers achieve such success in this sphere, and others fail. As a child can never be long quiet, he must be taken as quickly as possible; such portraits therefore can only be taken in fine weather.

The same remark applies to groups of many persons. No photographer has twenty or thirty head-rests at his disposal, he is therefore frequently reduced to the neces-sity of trusting to the good-will of his sitters for not moving. Those groups are very ugly which show a row of persons sitting beside each other like so many pagodas. The clever photographer therefore fixes the attention of his sitters by some occupation, such as looking at an album, eating, drinking, or card-playing. These persons must then of necessity assume various positions, some showing their front face, others their profile. The differences of complexion and dress in groups present great difficulties. Thus the exposure which suits fair faces will be too short for the dark. But as all must have the same length of exposure, it is not surprising that many parts of the picture appear under and others over exposed.

From these causes, no one can expect to appear to so great advantage in a group as in a separate portrait; if it happens so, this must be ascribed to chance.

It is usual for people to expect too much or too little from photographic groups. Your companion in the group is commonly well satisfied if his own portrait suits his fancy, quite forgetting the ungraceful arrangement of the rest, or the want of sharpness of his neighbours, who do not interest him so much.

Gentlemen ought to wear dark clothes. Light trousers and white waistcoats often appear in pictures as white patches, destroying the effect of the whole, for the principal light ought not to be concentrated on such accessories but on the head. Ladies, in choosing their toilettes, generally overlook the abnormal action of colours. On the occasion of the triumphal entry of the Emperor and the German army into Berlin, in 1871, the young ladies chosen to grace the ceremony were afterwards photographed in their white dresses trimmed with blue, and were not a little surprised that, in the picture, the blue trimming was as white as the dress. Blue often becomes white in photography, though there is an exception in the case of the blue uniform of the Prussian infantry. On the other hand, yellow and buff become black; the same remark applies to red. The photographer can, by a careful treatment of the negative, in some degree counteract this defect in case the clothes are of uniform colour.

The many-coloured toilettes, however, now in fashion, produce a disastrous effect. Materials whose beauty consists in variety of colour, it is evident, cannot make the same impression in photography as in nature.

Persons of dark complexion, also those who are stout, should prefer dark clothes. It is well known that white clothing increases in appearance the size of the figure. Thin and pale persons are advised, on the contrary, to wear light clothes, as a pale complexion would appear even paler when contrasted with black. Light clothing is always to be recommended for children. Materials should be chosen which by their lustre make a rich and picturesque impression; for example, satin, ribbed silk, taffeta, and silky materials. Woollen stuffs appear for the most part dull and lustreless, but they give very good effects as drapery. Persons with short and thick necks would do well to avoid high shirt collars, which make the neck appear still shorter. Ladies with similar attributes must lay aside velvet, ribbons, and such things around the neck; while persons of long necks will be improved by such ornaments.

The weather, the season, and the time of day present serious difficulties in photography. The days in winter being considerably shorter and darker than those in summer, commissions at Christmas are very inconvenient. Rainy days in winter are for the most part useless for photography; in summer they are light enough. The hours immediately before and after noon are the most favourable, as we have already stated in our chapter on optics.

Besides the clearness of the weather, the amount of light admitted by the instrument is an important matter.

The brighter the image produced by the lens is, the shorter will be the time necessary for a sitting. The greater the diameter and the shorter the focal length of a lens is, the more brilliant is its image. But it is by no means possible to increase the diameter or diminish the focal distance as much as you please, for optical defects in the lenses, which have not yet been overcome, stand in the way of this. The best lenses hitherto constructed (portrait lenses) only produce small pictures of cartes de visite, or at most of cabinet size. Larger pictures can be produced only by lenses giving weaker images; therefore they require longer sittings—a circumstance that makes the operation more difficult in cloudy weather or with restless models (as children) than the preparation of smaller pictures.

Accordingly, the latter show, on the whole, a greater technical perfection. As they are also very cheap, it is natural that the small cartes de visite, introduced by Disderi at Paris, have attained a general popularity, and given rise to a new kind of album, displaying the portraits of friends instead of poetry, and almost entirely supplanting the old scrap-book.

We can only approve of the modern album with its light-drawn likenesses of those we love or respect. We do not however like to see cartes de visite hung from the wall, they are too small to have any effect, and their frames are too insignificant.

Photography, like engraving, is an art that succeeds best on a small scale. Pictures of more than quarter life size cannot easily be taken from nature, but life-size pictures are demanded by the public. The photographer prepares these from a small negative, by help of the enlarging apparatus previously described in the chapter on optics.

For these pictures he requires a bright sun which, unfortunately, in our climate leaves him frequently in the

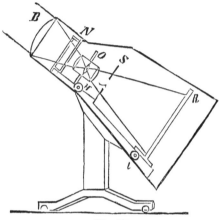

Fig. 67.

lurch. The small negative is placed in the apparatus (Fig. 67) at N, and on the table at R a sheet of sensitive paper is stretched. The lens at O projects a magnified picture of the little negative on the screen R, and when the apparatus is turned to the sun the large condenser B concentrates sufficient light on the picture to occasion a rapid browning of the paper: indeed, under favourable circumstances, a life-size copy can be taken in fifteen minutes.

Much has been said of instantaneous pictures. The Deputy Faucher remarked once, in the Prussian House of Deputies, "Instantaneous pictures are now possible. Portraits can be stolen by this process, and it will per-

haps be necessary to guard against it by the most extraordinary precautions—perhaps even masks will have to
be worn." This statement is based on a mystification.
Faucher had been made the victim of one of those photographers who, by incredible boasting and by puffing
themselves, seek to impose on the public. Instantaneous
pictures are possible if the object is clearly illumined by
the sun ; therefore an instantaneous picture of a brightly
illumined landscape can be prepared. A portrait in a
studio is quite another matter. Direct sunlight would
produce an unpleasant glare and sharp shadows upon a
dark background ; the eyes would be contracted, and a
very ugly picture would be the result. As we before
remarked, very powerful lenses have been constructed,
which admit of shortening the time of exposure. These
are, however, only suitable for very small pictures, and
are only employed for small, restless objects, like children,
in which case the photographer is satisfied if he gets the
chief part—that is, the head—as quickly as possible
into the picture.

It occurs frequently that persons wish to have an old
photographic picture copied. We here remark that,
though such a copy is possible, a photograph taken from
a photograph is never so fine as the original picture.
The cause of this is, on the one hand, the brown tone
of the photograph, which possesses very little photographic activity; on the other hand, the paper is sometimes glazed, and then produces a false light upon the
reproduced picture, or is rough, and then the fibre of the
paper gives the picture a disagreeable, coarse-grained
appearance. On that account it is easy, even for un-
practised eyes, to distinguish the copy from the original

photograph. Such copies are frequently to be met with at fairs and in stationers' shops, and can be bought for a ridiculously small price. In most countries, however, copying from original photographs is forbidden as piracy, and this prohibition ought soon to be introduced into Germany.

It has, indeed, been remarked that such infringement of copyright is advantageous to the public by lowering the price of favourite pictures.* But this advantage is no compensation for the injury thus inflicted on the author of the original pictures, who may have been put to considerable expense in obtaining his photographs, perhaps from distant places such as the Hartz mountains, or by the many failures before he has succeeded in making an imposing picture of a distinguished person. A great undertaking of this kind is seldom successful at first, and if his production is not protected by the law, he will prefer to give up producing such original pictures.

SECTION II.—LANDSCAPE PHOTOGRAPHY.

Its Scope—Difficulty of Taking Landscapes—The Photographic Tent— Application of Landscape Photography to Geography—Dry Plates— Stereoscopic Landscapes—Transparent Stereoscopes—Panoramic Picture

Landscape photography is a branch much less pursued than portrait photography. While portraits are generally made to order, it is a very rare thing to receive orders for a landscape. Such views are left to speculators, who employ photography as a means of representing favourite localities in largely frequented countries,

---

* Precisely the same argument can be advanced to defend the piracy of books, which is forbidden.

making thereby a profit with tourists. Thus photographers
wander through the noteworthy sights of our capitals and
mountains, and as the originals are accessible to every
one, each of the competitors tries to outdo his fellows
by excellence of work or cheapness. The copyist preys
upon these original photographers ; he does not undertake
any costly journeys, but awaits the issue of original
photographs to copy them at once, and offer them at a
low price. The inclination of the public is here favourable to the cheap seller. A landscape is seldom bought
for its value as a work of art, but more as a souvenir
of a happy hour, or as a reminiscence which, in subsequent years, will recall some interesting object, whether
a statue or a castle ; therefore less is demanded in the
case of views of landscape or architecture, and this is the
reason why landscape photography is not at present in a
very high state of perfection. The English are relatively
the best in this branch, because they get good prices for
their pictures and are protected against piracy. The
Swiss views of Mr. England have a world-wide celebrity :
in Germany pictures of equal merit are only produced by
Baldi and Würthle at Salzburg. Braun of Dornach also
deserves an honourable mention, having produced excellent landscapes, his Swiss views being known everywhere.

Superficial observers entertain the belief that one
photograph of a landscape must be as good as any other,
as the object remains always the same, and all are prepared by the same process.

Both assumptions are, however, erroneous. The object
is not always the same, for a landscape appears under
very different aspects in the morning and evening light,

or in fine and clear weather.   Whoever studies these
effects of light will soon discover at what hour a landscape will look most beautiful, and will choose it for
taking his view.   Accordingly, his picture will greatly
surpass that of a superficial and hasty photographer, who
takes the landscape as he finds it.   The choice of the
position is equally important.   A few feet higher or
lower, or a few paces to the right or left, and the whole
aspect of the scenery is altered (see page 144).   The man
having the eye of an artist, who knows how to seek the
best position, will at all times give the best picture.

The same remark applies to views of architecture
and sculpture.   It is evident that a photographer who
undertakes such views must be favoured by wind and
weather.   A breeze stirring the trees often injures his
picture, which may be rendered impossible for days
together by wind or rain.   To this difficulty may be also
allied that of an unmannerly class of people, who insist
on being taken with the picture, and thrust themselves
right in front of the photographic apparatus, making the
attempt impossible—a weakness that is more commonly
met with in Germany than elsewhere, and is the more
inexplicable as such people generally see nothing more of
the picture.

A great inconvenience for the landscape photographer
is that all the chemicals, bottles, glasses, dishes, etc.,
which are requisite for the process, must be carried on
the journey; nay, more, the photographer needs a trans-
portable dark room to prepare his sensitive plates.

The accompanying figure represents an apparatus of
this kind, together with the photographer at work.   The
upper part of his body only is in his tent, but the

space between him and the tent is made light-tight by suitable curtains. For the sake of facility of transport, everything in a dark tent of this kind is contracted into the smaller space. A yellow glass q lets light into the interior: the silver bath is in a box y, and the necessary water is in the cistern x, from which a pipe passes into the interior. The whole tent can be folded up, and forms a box of the size of z in the figure.

Fig. 68.

Though these arrangements are very compact, they are still of considerable weight, making the ascent of diffi- cult places, such as the Finsteraarhorn, the Wetterhorn, and the Jungfrau, impossible.

Dry plates are very useful in producing views of this kind, as they can be prepared at home and taken on a journey. The traveller can then dispense with a dark tent, collodion, silver bath, and the water for washing. The dry plates and camera are all that is necessary. It will be remembered from the description already given, that dry plates are prepared by washing an ordinary sensitized collodion plate, pouring on it some substance absorbing iodine, such as tannin, and then leaving it to dry. Unfortunately, plates prepared in this manner are less sensitive than the fresh wet plates, and the pictures they give appear less delicate than those taken with wet plates; the results, moreover, are very uncertain. Many plates spoil after a lapse of time. Then, the view obtained cannot be judged till it has been at home; and if, as often happens, it is a failure, it cannot be remedied by taking another picture. For these reasons, the wet-plate process has been preferred in taking landscapes, notwithstanding its inconvenience, and but few photographers work with dry plates.

Stereoscopic views of landscapes are very popular, for though so small, they surpass larger pictures in their power of truthfully representing landscapes. We have already described how these views are taken. If the light is bright and the lenses large, instantaneous views can be taken with the stereoscopic apparatus, and have been offered largely for sale.

The transparent stereoscopic views on glass, prepared by Ferrier and Soulier, are wonderfully beautiful. They are produced on a collodion film by placing the glass negative taken from nature on a dry plate, and then exposing it. A print of the negative is thus obtained upon the glass plate exactly in the same manner as with sensitive paper, only the action of the light is at first invisible, and must be developed with pyrogallic acid. As the production of such glass positives requires a more laborious and lengthened treatment than the paper pictures, their price is higher.

Latterly, however, by the help of a printing process (the Woodbury process), it has become possible to produce these pictures at a considerably cheaper rate. We shall describe this process further on.

Though at first sight landscape photography may appear unimportant, yet it is of the greatest use for geographical information. There is no better medium for conveying a true picture of foreign lands, of rocks, plants, and animal forms, than photography. It has even become an essential auxiliary of exploring expeditions, being alone capable of giving a perfectly faithful description of what has been seen. The inconvenience of transporting a photographic apparatus, and the ready decomposition of the chemicals, naturally limit the use of photography in exploring expeditions, and require a very expert photographer; but that these obstacles can be overcome is proved, among others, by the excellent views taken by Count Wilzek and Burger at Nova-Zembla, Baron Stillfried in Japan, Burger and Lyons in India, and Dr. G. Fritzsch in South Africa. We shall show in the following chapter the importance of landscape photographs for land-surveying.

Panoramic views are a special branch of landscape photography. The noted photographer Braun, of Dornach (Alsace), has sold for many years pictures which contained half the circumference of the panoramas of the Rigi.

Faulhorn, Pilatus, and other well-known points.   It is evidently impossible for a fixed camera to command at once such a panorama; the human eye cannot do this; the most we can survey is 90°, and this is only a quarter of the

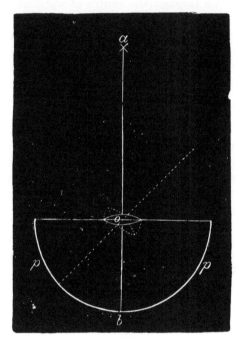

Fig. 69.

whole circumference.   If we wish to see the whole circumference, we are obliged to turn round.   Martens, a German engraver residing at Paris, conceived the idea of taking panoramic views with the help of a revolving camera, or

M

of a revolving lens in a camera. Let the reader imagine
a camera with a semi-cylindrical back, of which Fig. 69
represent a horizontal section, also a lens *o*. Then the
image of any point *a* is situated on the line *a o b*, which
is drawn from *a* through the centre of the lens. If the
lens revolves round its centre, the image remains immov-
ably at its place *b*; if it were to revolve about any other
point than its centre, the image would be displaced. It
is therefore evident that, if the lens revolves round its
centre, it can form successively an image of half the
horizon on the cylindrical surface. The only problem,
then, is to produce a sensitive cylindrical surface. This
is not difficult with sensitive paper, but much more
difficult with glass, which is extremely breakable in this
form. Accordingly, Brandon introduced a smooth plate,
which rolls as it were round the cylinder *p p*; that is,
which during the revolution of the lens is moved in such
a fashion that it always remains perpendicular to the
axis *o b* of the lens. The mechanism of such a camera is
rather complicated, but it has maintained its ground in
practice, and numerous panoramic views have been taken
with it. We must confine ourselves here to a general
description; those who seek for further details are referred to Vögel's "Manual of Photography."

SECTION III.—PHOTOGRAMMETRY, OR SURVEYING BY PHOTOGRAPHY.

Application of Photography to Measurement—Principle of Trigonometrical Measurement—Projection of Maps—Photographic Measurement
of Altitudes.

An essential difference between a photographic view
and an artist's painting is the fact, that it is not the
production of the operator's will, but that its outline and

design are subject to determinate laws.   All photographic views are produced by means of lenses.   A lens image of this kind is always in exact central perspective ; that is, each point lies on the straight line which can be drawn from the original through the optical centre of the lens. Let $a\ b\ c$ (Fig. 70) be three objects, K a camera (of which the outline is given to facilitate comprehension), and l its lens ; then the images of the objects $a\ b\ c$ will be on the lines $a\ o,\ b\ o,$ and $c\ o$—produced, viz., at $a'\ b'\ c'$; they have therefore the same relative position as the originals. Accordingly, a good photograph can serve to determine accurately the position of objects in nature, or to construct maps of the piece of ground that has been taken in the view.

For example, let the reader imagine the optical image of the camera placed flat upon the paper ; and then, in the middle of the picture (in this case the tree b'), a perpendicular line be drawn equal to the focal length $o\ b'$. It is only necessary to draw lines from the point o to the images of the objects $a\ c$ and F to at once find the directions in which the tower, the flag, and the trees will be seen from the position P.   If now a second view be taken from a point P' which lies in the direction of the flag F, a second view is obtained $c''\ b''\ a''$, which naturally looks quite different from the first in consequence of the change of position.   If this view from the second position be also placed on paper, in a position corresponding to P', and a line $b''\ o$, equal in length to the focal distance, be drawn as before, then the lines $c''\ o$ and $a''\ o$ will give the direction of the objects $a\ b\ c,$ as seen from P'.   If these lines be sufficiently produced on the paper, they will intersect at points the situation of which corresponds exactly to

that of the object; and thus, in two views from two
points, the means is afforded of constructing a map in
which the situation of all points contained in both views
is exactly given.

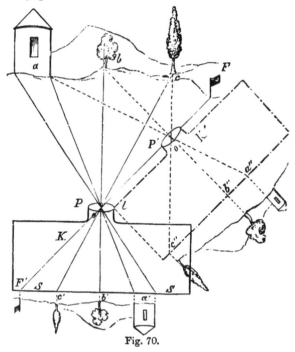

Fig. 70.

A different procedure is followed in ordinary trigonometrical measurements. The first step would be to
measure the distance $P P'$, then to set up an instrument
for measuring angles at P, and to determine the angle
made by the line $P P'$ with the lines $a o, b o, c o;$ the

same operation being repeated at the other end of what is called the base line $P\ P'$. It is evident that so many measurements must be made at both points as there are objects of interest, whereas a photographic picture in one operation fixes the relative position of all objects. Accordingly, there is a considerable economy of time in the application of photography; and this is of great moment in time of war, when frequently, in consequence of interruptions on the part of the enemy, the leisure is wanted which is necessary for triangulation; also in journeys, when the stay at particular places is too short to make observations requiring time.

Therefore this process has great advantages in exploring expeditions, whose photographs would thus be doubly valuable; for not only do they give a view of the country, but also data for the projection of maps. For this purpose two views taken from each end of a base line are necessary. Moreover, the taking of these views must be carried out with mathematical accuracy: the camera must be placed in a perfectly horizontal position; its lens must give a perfectly correct image; the plates must be absolutely level, etc.—conditions which are not easy to comply with. To these other difficulties are added, proceeding from the very nature of photography, which requires clear, bright weather; with a troubled sky, or when the atmosphere is misty—the aerial perspective of landscape painters—distant objects are often so indistinctly given in the picture that no correct measurement can be made of them, though the surveyor can clearly distinguish and measure from nature in such weather. Further, the direct action of the sun offers difficulties to photography. If it stands in front of the camera and shines full on the lens, it often occasions fogging on the plate, greatly affecting the value of the picture for purposes of measurement. All these circum-

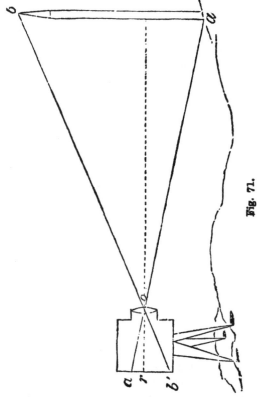

Fig. 71.

stances militate against the application of photogrammetry, as this mode of measurement has been called by Meydenbauer, who has long used it. Meydenbauer pre-

pared a good map of the Unstrutthal by this method.
But the experiences during the campaign of 1870
are not so satisfactory (the royal Prussian staff tried
the process before Strasburg)—perhaps the imperfection
of the apparatus used occasioned the unsatisfactory
results.   It is to be hoped, in the interests of geography,
that future attempts will succeed in making this important method practically available.

Photography can determine the elevation of mountains and of buildings, as well as determine positions in
a plain.   Let it be supposed that $a\ b$ (Fig. 71) is a tower,
and $a'\ b'$ its image in a photographic camera.   It is
evident that the image will be smaller than the object.
According to a well-known mathematical law, the size
of the image $a'\ b'$ is to that of the tower $a\ b$ as the
distance of the image from the lens $o\ r$ to the distance of
the tower from the lens.   This gives the proportion:

$$o\ r : D = a'\ b' : x,$$

where D is the distance of the tower from the camera,
which can be measured.   The height of the tower can be
easily found from the above proportion.

Meydenbauer has deduced the dimensions of the
ground-plan and elevation of a house from a photograph.

Section IV.—Astronomical Photography.

Its Application—The Photographic Telescope—Eclipses—Protuberances
—Corona—Sun-spots—Enlarged Images of the Sun—Researches of
Rutherford—Astral-Photography—Pictures of the Moon—Photo
graphs of the Spectrum—Photography and the Transit of Venus.

The application of photography to astronomy is twofold.   First to give a faithful representation of certain

phenomena of the heavens—which change so rapidly that
the draughtsman cannot follow them; for example, the
phenomena of eclipses, or others which are inconvenient
to draw, such as sun-spots. Secondly, astronomical photography has to produce pictures of heavenly bodies which
can be used for measurements. Photography has been
used successfully for both purposes, and it is employed
daily as an auxiliary to produce views of sun-spots at
several observatories; for instance, in Germany, at the
observatory of Herr Von Bülow, Privy Councillor at
Bothkamp, near Kiel.

The mode of preparing astronomical pictures differs
little from that of ordinary photographs. An ordinary
photographic apparatus could be used for this purpose,

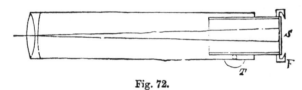

Fig. 72.

were it not that it gives too minute images of very remote
objects, such as the stars. The size of the picture is in
direct proportion to the focal length of the lens; therefore, in taking astronomical photographs, lenses are used
the focal length of which is very long, by converting an
astronomical telescope into a photographic camera.

The accompanying figure shows a telescope of this
kind adapted to photographic purposes. The objective
O remains in its place, the eye-piece, which would be
fixed at the other end of the tube, is taken away and
an apparatus V (Fig. 72) is substituted for it, which is

identical with the hinder part of a photographic camera. It contains a ground glass slide *S*, which, after the image

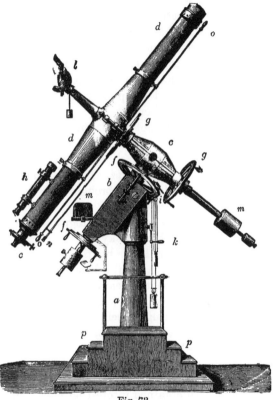

Fig. 73.

has been focussed, can be exchanged for a sensitive plate. The focussing is effected by a rackwork motion at *T*.

But now a difficulty occurs through the motion of

the stars, which necessitates the telescope following this movement. For this purpose, the stand of the telescope is furnished with a driving-clock, which causes it to revolve in the direction of the course of the stars, so that the telescope is what is called equatorially mounted. Fig. 73 shows an arrangement of this kind.

The oblique support of the telescope resting on the foot is parallel to the axis of the earth. The polar axis of the telescope is turned completely round in this support once in twenty-four hours.

The telescope d d is not fixed immediately on this axis, but on an axis c, at right angles to it; it can be turned round the latter (the declination axis) in all directions perpendicular to the axis c i. The movement of the two axes allows any star to be brought into the field of the telescope.

The first attempt to employ photography for astronomy was made by Berkowsky, at the Königsberg Observatory, in the year 1851, by the help of Bessel's noted heliometer, during a total eclipse of the sun. He obtained a daguerreotype, the beauty of which was much lauded, and which showed very well the remarkable phenomena that appear during an eclipse of the sun—flame-like formations that stand out from the darkened disc of the sun, called protuberances. In the year 1860, Warren de la Rue in England, and Secchi at Rome, undertook an expedition to Rivabellosa, in Spain, to observe the eclipse of the sun, and both obtained interesting views on collodion plates. In 1868, the Government of the North German Confederation equipped an expedition to observe the eclipse of August 18th, and sent Dr. Fritzsch, Messrs. Zencker, Tiele, and the author to take photographs.

Another photographic expedition was sent out by the English Government to India. Besides these, the German, English, Austrian, and French Governments sent out expeditions for the ocular observation of the phenomenon.

Obstacles were, no doubt, encountered by these expeditions, nevertheless they produced results that finally settled the question about the nature of the protuberances, and moreover gained experience that materially lessened the labour of subsequent photographic observers.

We introduce a description of the expedition to Aden, giving a faithful account of the obstacles associated with an undertaking apparently so simple. The author wrote from Aden the following account of his arrival and residence at that place :—

" The aspect of Aden is by no means cheerful. A broken mass of rock, bare and desolate, the remains of an extinct volcano, among which are warehouses, shops, coal-sheds, flag-staffs, etc. ;—such was the appearance of the place that was to be our residence for a fortnight. The colour green was entirely wanting there.

" Our luggage and ourselves were conveyed to land amid the shouts, quarrelling, and tumult of the Arab mob. On landing we learned that our colleagues, who had preceded us, had been received in the most courteous manner by the British authorities, and that two Indian huts—bungalows—used in this climate on the east side of the peninsula, had been assigned us as our station.

" After a long search we found the locality and our comrades, together with the members of the Austrian expedition, Dr. Weiss and Messrs. Oppolzer and Riha, and in as excellent quarters as could be wished on this desert coast. The English authorities acted the part of host in the most generous manner. A whole staff of servants, cook, etc., waited upon us; carriages, camels, and donkeys were at our disposal, and all our wishes were anticipated. Thus our bodily comfort had little to desire; the temperature (26° Reaumur *), might be called low for the Red Sea, for a fresh breeze was always blowing on the heights of the Marshagill, on which our bungalows stood, and contributed greatly to our comfort.

"There still remained ten days to prepare for taking views of the eclipse. They were employed in preparing stands, mounting and arranging our telescopes. We used as observatory a bungalow, which we partly unroofed, in order to use our telescopes, and we converted the rest of its interior into a laboratory, washing-room, and store-room. In this bamboo cage—for it was nothing more—we were tolerably protected from the wind, but less so from the dust. Water was brought to us by donkeys, in leather bags. Two tents that we had brought from Europe answered the purpose of dark rooms. Spare apparatus for taking landscapes and portraits, that we had brought with us, gave us the material for taking views of the country and its population, and were also a useful means of testing our chemicals.

"Some trifling defects in the latter were quickly remedied, but it was not so easy to remove the effects of the dust and human exhalations. During the slightest exertion in that damp atmosphere, the perspiration flowed from the body in streams: it ran from

* 90·5° Fahrenheit:

the tips of the fingers, dropped from the face, and often a well-cleansed or prepared plate was spoilt by a drop of sweat. Nevertheless, practice taught us how to encounter this obstacle; some attempts at taking the sun, etc., were successful; and we were able to tranquilly await the eclipse. Only one thing gave us serious uneasiness—that was the weather. All accounts of Aden had unanimously represented its sky as perfectly clear, competent witnesses having asseverated that it rained there at most three times a year, and that clouds were exceptions.

"We were therefore not a little surprised when, on our arrival, we found the volcanic heights of Aden quite concealed with clouds, and when we were greeted with a shower of rain on the following morning. But we became still more anxious when, day after day, the sun was concealed by clouds, and this weather became worse rather than better in the course of time. The prospect of succeeding in our main object looked dreary enough, and soon all our hopes deserted us.

"On the day of the eclipse we left our quarters about 4 a.m. Nine-tenths of the sky were cloudy. We set to work in a resigned frame of mind. It was the undertaking of the North German expedition to photograph the eclipse throughout its continuance. For this purpose we used a telescope with a six-inch achromatic lens, with a focal length of six feet. This lens, made by Steinheil, gave an image of the sun three-fourths of an inch in diameter, which could be taken on a photographic plate by the help of an ordinary dark slide containing a plate for two pictures. As the sun and moon move, such an instrument, if stationary, would only give ill-defined views. Accordingly, the telescope was connected with clockwork, which gave it a movement corresponding with that of the heavenly bodies. To avoid all agitation of the telescope, the closing lid of the objective was not placed directly upon it, but on a separate stand, and was connected with the telescope by an elastic sleeve.

" The duration of the total eclipse was at Aden three minutes, in India five. We had, however, chosen our station at Aden because photographic observers were already present in India, and because the eclipse began about an hour sooner at Aden than in India. Thus, by comparing our observations with those in India, a judgment might be formed whether those wonderful phenomena of light called protuberances, seen during total eclipses, changed or did not change in the course of this time. It was our present endeavour to take as many views as possible of the phenomenon in three minutes. With this object we had regularly practised, as artillerymen do their cannon.

" Dr. Fritzsch prepared the plates in the first tent, Dr. Zencker pushed the slides into the telescope, Dr. Tiele exposed them, and I developed them in the second tent.

"We had ascertained that it was possible in this manner to take six views in three minutes.

" The decisive moment approached. The cloudy sky anxiously surveyed by us, showed to our great satisfaction a few breaks through which the disc of the sun, partly concealed by the moon, and appearing as a crescent, was visible. The landscape appeared in the strangest light, being almost a half-and-half mixture of sunlight and moonlight. The chemical action of the light showed itself very weak. An experimental plate only gave a view of the clouds after fifteen seconds' exposure. The sun's crescent became gradually smaller, the break in the clouds gradually increased, and we took heart.

"The last minutes preceding the total eclipse, which occurred at 6.20, fled on wings. Dr. Fritzsch and I crept hastily into our tents and remained there, preparing plates and developing. The consequence was that neither of us saw anything of the total eclipse. Our labour begun, the first plate was exposed, as an experiment, from five to ten seconds, in order to see what was the proper time for exposure.

"Mohammed, our black servant, brought the first slide into the tent to me. I poured the iron developer upon the plate, waiting breathlessly to see the result. At this moment my lamp went out. 'Light! Light!' I exclaimed; but no one heard me—every one had enough to do. I stretched my right hand out of the tent, holding the plate with the left, and fortunately grasped a small oil lamp, which I had placed ready for all emergencies, and now I saw the image of the sun appear upon the plate. The dark rim of the sun was surrounded by a series of peculiar prominences on one side, while on the other side appeared a singular horn—both phenomena perfectly analogous in both pictures. My delight was great, but there was no time for rejoicing; soon the second plate and, a minute later, the third plate were in my tent. The sun is emerging,' exclaimed Zencker. The total eclipse was over; but all this appeared as the work of a moment, so quickly had the time passed. The second plate showed under development only faint traces of an image, a passing veil of clouds had almost destroyed the photographic action at the moment of exposure. The third plate showed again two successful views with protuberances on the outer rim.

"Rejoicing in this success, the plates were washed, fixed, varnished, and at once, although with very imperfect instruments, copies were taken on glass, and these, to avoid accident, were transmitted separately to Europe.

"Our extraordinary good fortune is apparent from the fact that, at a place distant only half a league from our station, nothing was seen of the total eclipse on account of the veil of clouds.

"We did not stay long at Aden after our chief object had been attained: in three days the steamer proceeded to Suez. Telescope, clockwork, and our heap of instruments and chemicals were quickly packed, placed on camels and conveyed to the harbour. On the 21st of August we bade adieu to the barren, rocky island, and started for Suez."

Aden was one of the points where the eclipse was first visible. As previously stated, England had also equipped a photographic expedition, which stationed itself at Guntoor in India. The eclipse was observed an hour later in India than at Aden. The same protuberances appeared in the Indian photographs as in those at Aden, but they present a very different form, which seems to show that these prominences are not compact bodies, but formations of a cloud-like nature; and this supposition was converted into certainty by Jansen's observations with the spectroscope, made simultaneously.

Jansen showed that in a total eclipse the protuberances display bright lines in the spectroscope. As this is the property of gaseous bodies only, the question as to the nature of the protuberances was solved. Jansen determined at the same time the exact position of these bright lines, and discovered that the gaseous substance was glowing hot hydrogen. He subsequently made the discovery that an eclipse was by no means necessary in

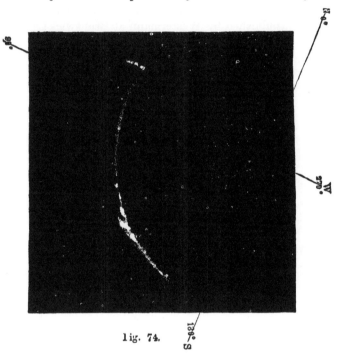

Fig. 74.

order to detect the bright lines of the protuberances. They are seen on clear days if the slit of a spectroscope be directed on the sun's rim, and the changeable nature of these protuberances can every day be observed by the appearance and disappearance of these bright lines.

N

Zöllner of Leipzig even detected with the spectroscope a sudden burst of gas, also the sudden breaking away of gas clouds from their substratum, and their dispersion, all in the space of a few minutes.

We add a faithful copy of the Aden photographs, taken from Herr Schellen's excellent work on spectral analysis (published by Westermann at Brunswick).   The

W

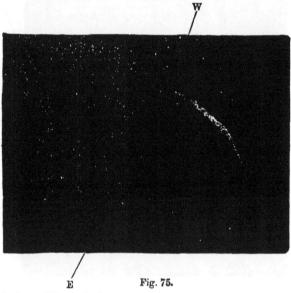

E   Fig. 75.

first view (Fig. 74) gives us the eastern rim of the sun; the western was covered by clouds.   It is easy to recognize in it the large horn-like protuberance, which has an elevation of 184,000 miles, and gives an idea with what immense force masses of gas are projected from the surface of the sun.   It shows, further, the remarkable protuberance to the left, in which the masses of gas appear like powerful jets of flame driven sideways by a tempestuous wind; a light region surrounding the protuberances forms the glowing hot stratum of vapour permanently surrounding the rim, named chromosphere.

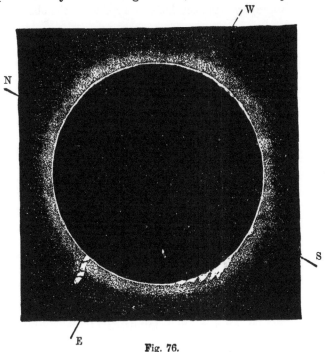

Fig. 76.

The second view presents only a series of point-like protuberances on the western rim of the sun, but these points are so large that our earth could almost find room in them. The eastern part of the sun was under the clouds during the taking of this view.

Finally, the third view gives a perfect representation of the total eclipse as it was observed in India. Besides the protuberances seen at Aden, there is another on the western rim of the sun, which was quite covered by clouds at Aden.

Photography has been latterly applied to the observa-

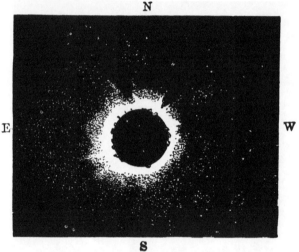

Fig. 77.

tion of total eclipses on a more magnificent scale. Thus, on the 7th of August, 1869, hundreds of photographers were actively employed in observing the total eclipse of the sun at Iowa, in North America, and more than thirty telescopes were set up to retain the phenomenon. By these observations, the question respecting the nature of the protuberances was finally set at rest, and the only question that remained related to the corona. By corona is meant a kind of nimbus of white light surrounding the sun when totally eclipsed.   Many observations of total eclipses have been undertaken to decide its nature. A very beautiful view of the corona was obtained by Whipple, at Shelbyville in Kentucky, August 7th, 1869. The feeble light of this phenomenon renders a much longer exposure necessary than in the case of the protuberances.   At Shelbyville, the exposure for the corona lasted forty-two seconds, whereas five seconds sufficed to take the protuberances.   Nor was the nature of the corona as yet determined.

In 1870 an English expedition, conducted by Lockyer, was sent to Catania to observe the corona, and the author accompanied it.   Unfortunately, owing to the unfavourable weather, the observations were only partially successful.   However, a detachment of the expedition, conducted by Brother, in Syracuse, succeeded in obtaining a good view of the corona, and we give a faithful woodcut (Fig. 77) copied from this photograph.

The black prominences round the sun's disc give the situation of the protuberances which were visible on the day of the eclipse.   We call attention to the fact that they are not visible in the photograph of the corona. To take a view of the corona requires an exposure eight times as long as for the protuberances.   During this long exposure the images of the protuberances received too much light, and have therefore become paler instead of brighter, so that their outline becomes confounded with the less bright parts.

Photography is applied to other important purposes in astronomy besides taking pictures of eclipses.   Views of the sun are taken daily.   The observation of centuries has established that the sun is continually changing : spots appear, increase, and disappear. All these phenomena were at an earlier date explained as openings in the cloudy luminous atmosphere of the sun, which was supposed to surround its dark central mass. Now they are looked upon as immense whirlwinds, which rage in the atmosphere of the sun (see Schellen's "Spectral Analysis," page 200), or as cloud-like condensations. Their nature has not been perfectly ascertained. These sun-spots follow the revolution of the sun's body round its axis, and experience manifold changes during this time. It is only by means of these spots that the duration of the sun's revolution has been determined. Recent observations have established that the number of the spots increases and decreases periodically, and that this period is connected with the magnetic phenomena of our earth. These circumstances have led to a closer study of the spot formations, and photography has offered a valuable aid for this purpose. It gives at a particular moment a faithful view of the sun's surface, and photographs taken daily give us the most exact representation of the spots, their size and number. A comparison of the views during one month affords an instructive survey of the changes on the sun's surface, relating more faithfully than words the history of the central body of our planetary system. Lewis Rutherford, of New York, who has made valuable contributions to astronomical photography, has taken a great number of these views at the photographic observatory built at his own expense.

These views, taken on successive days, exhibit manifold groups of spots, often of considerable size ; and the change in their form and position is thus accurately recorded (the latter consequent upon the revolution of the sun). These impressions are not prepared, as were the pictures of the eclipse, in the principal focus of the telescope, but in an additional apparatus (Fig. 78) which

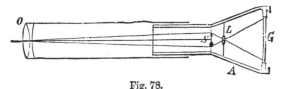

Fig. 78.

answers the purpose of a magnifying apparatus. This contains a small lens L, which projects on the ground glass G an enlarged image of the small representation of the sun S produced by the great lens O.

In this manner Rutherford obtains directly a picture of the sun about two inches in diameter. This enlarging apparatus is not to be recommended in the case of eclipses; for the brightness of the image produced by the great telescopic lens is materially weakened by the enlargement. When the image is twice the size, the weakening of the light is fourfold; when it is three times as great, the weakening is ninefold, and so on. In views of the unclouded sun this is of no consequence, for its light is so intense that it bears a considerable enlargement, and yet remains bright enough to give a view on a momentary exposure. The protuberances, however, give out much less light, and if their image were magnified they would become so faint that a longer exposure would be required than the duration of the eclipse.

The solution of other important astronomical problems has been attempted with the help of photography; for example, the production of pictures of the stars.

The object of such pictures is the representation of the constellations, or the relative position of the stars. It always has been one of the principal objects of astronomy to determine the position of the fixed stars. It may be thought that the catalogue of the stars is already completed, and that the matter is settled; but this is not the case. As far as photography can at present be applied, that is, to stars of the ninth magnitude, the catalogue is not complete. Moreover, the measurements of the past may require correction in consequence of improved methods.

The photographic process is of great importance for this purpose, because it offers advantages in the facility of its application and correctness of its results. Many readers may inquire why so much trouble is taken to determine with the greatest accuracy the positions of thousands and millions of fixed stars. The answer is that the fixed stars are not stationary as their name implies; nothing is stationary in nature, and hence such study is never at an end. No doubt the fixed stars change their position so slowly that the builders of the pyramids four thousand years ago beheld the constellations much as we do. It is only the finest astronomical measurements that show any change within a limited number of years. However, the study of the proper movement of the fixed stars has now begun, and requires very accurate measurements carried on for generations.

Another interesting point comes into consideration in connection with this subject. On the one hand, the fixed stars are not without movement; on the other, their distances from the earth are very various, the nearest being enormously great. The photographer who wishes to take a graphic view of an object, will always try to take it from different points. Two views of a moderately remote object taken from two points only two inches apart, appear different to the eye, and produce, when viewed in a special manner, a stereoscopic effect. No distance on earth is great enough to give different pictures of the same constellation; nevertheless, within the space of one year we describe a circle round the sun having a diameter of 184 millions of miles, so that in half a year we are 184 millions of miles from our present position. This enormous distance is in certain cases just sufficient to show a change in the relative position of certain stars, though not to the naked eye. By this means the distance of the nearest fixed stars has been determined, amounting to billions of miles.

By careful comparative measurements of positions of neighbouring stars, continued for years and centuries, a change can be proved to exist, and the proper movement of the stars can be calculated. The distance of the stars can be deduced by carefully collating the yearly recurring changes in the positions of the stars. It is evident that photography, which affords the means of fixing these positions, must be of the greatest value for both these astronomical problems.

Photography of the stars was first introduced about twenty years ago by Professor Bond, of Cambridge, Massachusetts, but it was Mr. Lewis Rutherford, of New York, who perfected the method. He constructed a photographic objective of 11 inches diameter and about 13 feet focus. This objective shows a considerable focal difference; that is, the violet and blue rays have a different focus from the yellow and red. If a clear image of the star is focussed, the sensitive plate would be adjusted to the focus of the yellow rays, and the focus of the chemically operative blue rays would fall beyond the sensitive plate, and an indistinct picture would result. The plate must therefore be adjusted to the focus of the blue rays; but this is not so easy to find. After it has been found approximately, it is corrected by taking different photographs of a star, changing the position of the plate each time. The point is determined at which the best and sharpest picture is obtained, and, by continual repetitions of the attempts, the chemically active focus of a lens of 13 feet focal length can be accurately determined to within $\frac{1}{150}$ of an inch. It is well known that all heavenly bodies have the same focus, on account of their great distance.

No photographic objective gives a correct picture with a large surface. Accordingly, to obtain the accuracy required by astronomy, the surface to be devoted to the image can only be very small, or about $1\frac{1}{2}$ degrees in diameter. Any remaining distortions are controlled and corrected by photographing a very correct scale, and comparing the picture with the original. A field of $1\frac{1}{2}$ degrees, or three times the moon's diameter, embraces the well-known constellation of the Pleiades.

Rutherford's telescope is arranged as in Fig. 73 (p. 169); it is moved by clockwork, so as to exactly follow the movement of the stars.

The views of large stars taken with it, after a short exposure, all appear as small round points which can only be seen through a magnifying glass. In the case of a long exposure their size depends on the more or less strong vibrations of the atmosphere, which occasion the flickering of the stars. Stars of the ninth magnitude can be photographed with an exposure of eight minutes; the light of these stars is ten times weaker than that of the faintest which can be detected on a clear night by the naked eye, and their images are very small points. It would be difficult to distinguish these small points from dirt spots on the plate. To do this, Rutherford makes use of an ingenious device. He brings the telescope, after the first exposure of eight minutes, into a slightly different direction, and makes a second exposure of eight minutes, while the clockwork continues to operate, and moves the telescope correctly in this second direction.

In this manner two images are obtained of every star on the plate, closely adjacent; the distance and relative position being in all the same. These double images can be easily found on the plate and distinguished from accidental spots. If the telescope stops, it is evident that the images of the stars make a movement on the plate, the bright stars describing a line. This line is of great importance to determine the direction from east to west on the plate. For faint stars which leave no line a third exposure is necessary to determine this direction. This is done after the clockwork of the telescope has been stopped for some minutes.

Rutherford has already taken numerous views of the stars, and they will serve as important means of comparison, after the lapse of centuries, in order to discover what change has taken place in the position of the fixed stars.*

* Details respecting Rutherford's observatory are contained in the "Photographischen Mittheilungen," Jahrg. 1870. Berlin: Oppenheim.

But another heavenly body invites us specially to study it by the help of photography; that is, our nearest neighbour, the moon. The unassisted eye recognizes its uneven surface ("mountains in the moon") and the varying shades of its ground (moon spots). Its surface contains a thousand problems, appearing as a rigid, almost vitreous, waterless, airless waste.

Warren de la Rue has tried to take photographic pictures of this singular body, which is so near to our earth and yet so different; he succeeded in obtaining, with the help of a telescope, a small view of the moon, which he enlarged to 24 inches diameter with an enlarging apparatus (p. 94).

The moon gives out less light than the sun. It is therefore taken to the best advantage in the principal focus of the telescope. (See Fig. 72, p. 168.) In the most favourable case, three-quarters of a second suffices for exposure, but it is rare to obtain sharp negatives, owing to the disturbance of the atmosphere, and to obtain a sharp image of the moon is a test of patience. After Warren de la Rue, Rutherford is celebrated for his moon-pictures; his improved telescope, set up purposely for photographic purposes, gave a still sharper image of the moon than De la Rue's, and our frontispiece is a diminished copy of the enlarged picture of the moon from an original, for which we are indebted to Rutherford, forming a genuine map of the moon of no small importance to astronomy.

Some years ago Schmidt, at Athens, maintained that an extinct volcano observed by Mädler is no longer to be found, and he thereby proved the possibility of changes on the apparently rigid surface of the moon. If a photograph of the moon's surface could have been taken forty years ago, when Mädler observed the volcano, we should now be certain about this point, which is still doubtful.

But the sun and its eclipses, the moon, and stars are not the only objects of astronomical photography. Its province extends further since the discovery of spectrum analysis.

When it was discovered that the wonderful lines intersecting the sun's spectrum (see chap. vii. p. 62) were occasioned by glowing substances of different kinds, and that each element shows invariably the same lines, so that the presence of certain spectral lines establishes the presence of certain elements, it became necessary to possess an exact map of the countless lines of the solar spectrum. Then, by comparing this map with the spectrum of a flame or of a star, it could be at once seen what substances give these lines. Kirchhoff, one of the discoverers of spectrum analysis, and Angström have prepared such maps of the solar spectrum. Their labour would have been materially simplified had Rutherford published his photograph of the spectrum a year earlier.

Rutherford's photograph only shows the lines of the chemically active part of the spectrum—from green to violet—but it does this with wonderful clearness. Many lines that appear faint to the naked eye show themselves strong and sharp in the picture; nay, lines are discovered in the photograph of the spectrum which Kirchhoff did not see at all in the spectrum itself.

The causes of this phenomenon may be twofold: either the eye is not sensitive to certain rays of light—as we know it is not influenced by the ultra-violet rays,

which have a strong photographic effect,—or it is possible
that changes take place in the sun : that at certain times
fresh substances come to its surface, and thereby new
lines become apparent.

The photograph of the spectrum may be taken with
the aid of an ordinary spectroscope, seen in Fig. 79. This
consists of a tube *A*, which has a fine slit *F*, through
which the light penetrates, and at the other end a lens,
which makes all the rays from the slit parallel, and conducts them to the prism *P*. The rays are then refracted

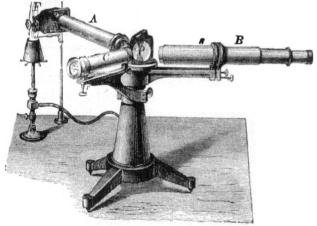

Fig. 79.

and pass into the tube *B*, and can be observed through its
narrower end. If the object is to photograph the spectrum thus seen, a photographic camera is placed light-
tight on the tube *B*, its eye-piece is drawn a little out,
and then the image of the spectrum appears upon the
ground-glass screen.

Attempts have been made to solve other important problems by the help of photography. Thus, Dr. Zencker hoped to be able to trace the path of shooting stars by means of it. Unfortunately, these were found to give out too little light to produce, while they lasted, an impression on the photographic plate.

The transit of Venus affords a new and grand problem for photography.

In determining the distance of heavenly bodies, the diameter of the earth's orbit is taken as a base line; therefore the knowledge of the exact length of this base is assumed. Now, this amount has only hitherto been determined by approximation, and is in round numbers one hundred and eighty millions of miles.

Many efforts have been made to determine this distance more accurately. It is, however, a problem of great difficulty. Let it be conceived that there are at two opposite points of the earth, a and b (Fig. 80), two observers who look with telescopes at a star x, and measure the angles which the rays $x\,a$, $x\,b$ make with the line $a\,b$; it can be determined from both angles and the line $a\,b$ (which is easily found) what the distance of the star is from a or b. This is the trigonometrical method, and it gives reliable results, if the distance of the star is not too great; thus, for example, the distance of the moon, which is about ten of the earth's diameters, is easily ascertained. If the star to be measured is too remote, the rays $a\,x$ and $b\,x$ are nearly parallel, no difference exists between the two

Fig. 80.

angles at a and b, and the trigonometrical method is useless. This is the case with the sun, which is ninety-five millions of miles from the earth. We can therefore only ascertain its distance by indirect methods.

According to a law discovered by the celebrated astronomer Kepler, the squares of the periods of revolution of the planets vary as the cubes of their distances from the sun. Thus, if the period of the earth's revolution is U, and that of Venus u, the distance of the earth E, that of Venus e, according to this law,

$$U^2 : u^2 = E^3 : e^3.$$

If the cube root is extracted from both we have—

$$\sqrt[3]{U^2} : \sqrt[3]{u^2} = E : e, \text{ hence,}$$

$$\sqrt[3]{U^2} - \sqrt[3]{u^2} : \sqrt[3]{u^2} = E - e : e.$$

But $E - e$ is the distance between the earth and Venus. When this has been determined, three terms of the proportion are known; for the periods of the revolutions of Venus and the earth are accurately known. Then, by simple rule of three, the fourth term e can be found; that is, the distance of Venus from the sun. If to this is added the distance of the earth from Venus, we obtain the distance of the earth from the sun, which was required.

Thus the determination of our distance from the sun depends on that of our distance from Venus, which must be taken at the moment when Venus is between the earth and the sun. But Venus is only visible at the moment when it is between the sun's disc and the earth. This is only exceptionally the case—twice in every century—and then it appears as a small black point on the sun's disc, which, however, continually changes place, on account of the earth's movement and its own.   This circumstance renders simultaneous measurements at two different and remote points of the earth very difficult, and therefore the idea has been entertained of using photography as an auxiliary.   If by its help, and in the manner described above, a sun picture is taken during the transit of Venus, the distance of Venus from the sun's centre can be easily measured upon it.   The centre of the sun is a fixed point which can be assumed to be stationary.

If the earth is supposed to be at E (Fig. 81), Venus

at V, and the sun at S, the observer at a will see Venus below the centre of the sun, while an observer at b will see Venus above it.   Accordingly, Venus will present a different position to the sun's centre on photographs taken at various points of the earth.

Now, the position of the centre of the sun is very accurately known.   The sun's diameter subtends an angle of about 30 minutes, so that if divided into 30 parts, each part would represent an angle of one minute.   It is only necessary therefore to measure how many of such parts Venus is distant from the sun's centre, to find at once the angular distance of Venus from this point or $m\,a\,V$.   If this angle is subtracted from the angle which the line $a\,b$ makes with the direction of the sun's centre $m\,a\,b$, the remainder will give the angle $V\,a\,b$ which Venus makes with the line $a\,b$,

Fig. 81.

which gives all the data necessary to calculate the

O

distance of Venus, and, from that, the distance of the central body which forms the foundation and base of all astronomical measurements.*

The determination of this angle by photography is of special value, as this measurement can easily be made at any convenient time, whereas direct measurements can only be made while the phenomenon is visible, and hence many errors are introduced in the excitement of the moment. Measurements of this kind require apparatus of the most accurate description, and the adoption of many precautions; therefore preliminary experiments have been already commenced to determine the degree of accuracy which a measurement by means of photography admits. If these preliminary experiments give a favourable result, numerous photographic expeditions will be sent out to observe the transit of Venus. Germany proposes to occupy five stations: Tschifu in China, Muscat on the Persian Gulf, Kerguelen's Land, and the Auckland Islands. Besides these, England, France, Russia, and America are equipping photographical expeditions which will occupy various points, and thus we may hope that though some stations may be visited by unfavourable weather, still numerous plates will be obtained by means of which this great astronomical problem can be solved.

* Our space does not permit us to enter into the details of the method of determining the solar parallax; it is only our purpose to give a plain intelligible statement of the first principles of the problem. Those who are specially interested in the subject are referred to " The Transit of Venus over the Sun," by Dr. Schorr. Brunswick: Vieweg. 1873.

SECTION V.—THE APPLICATION OF PHOTOGRAPHY TO THE OBSERVATION
OF SCIENTIFIC INSTRUMENTS.

The Thermometer and Barometer—Neumeyer's Apparatus for
determining the Depth of the Sea.

Meteorological observations require a daily reading of
the barometer and thermometer.   To economize this
reading, and yet to receive a perfectly safe register of the
state of the thermometer and barometer at each minute,
photography has been turned to account.   Let the reader
imagine behind the tube of a thermometer R (Fig. 82)
or barometer a drum, which revolves round its axis a
by means of clockwork.   Let sensitive
paper be wrapped round this drum,
and the whole be enclosed in a cylinder
S, which has only a small slit behind
the thermometer through which the
light can penetrate.   The upper part
of the thermometer will let the light
through, while the thread of quicksilver will stop the light.   Therefore
the strip of paper above the quicksilver

Fig. 82.

will blacken, and the limit of the blackening on the
paper will rise and fall with the mercury.   Now the time
can be marked beforehand on the paper.   As the drum
revolves once in twenty-four hours, the strip of paper
need only be divided perpendicularly into twenty-four
parts, and the first part be moved opposite the thermometer directly the clock strikes twelve, after which

the whole may be allowed to revolve. Thus the blackened strip will show the height of the thermometer at all times of the day. In the same manner, the height of the barometer can be registered by photography.

Professor Neumeyer has latterly employed a similar instrument to determine the temperature in the depths of the sea. As there is no light producing chemical action at those depths, Neumeyer sends down a light-producing apparatus. This consists of a galvanic battery, and a Giesler's tube; that is, a tube in which very attenuated nitrogen gas is enclosed, and through which the electric current is passed. The tube then gives out a faint light. But this faint light has a powerful chemical action, because it contains many of the invisible ultra-violet rays (see p. 64), and in three minutes it effects the blackening of the paper. Neumeyer also attempts to determine with his apparatus the direction of the oceanic currents. For this purpose the apparatus has attached a vane like that of a weathercock, which can move the instrument in any direction whilst suspended. If any currents exist, they will turn the apparatus so that the vane is parallel to their direction. A magnetic needle is enclosed in the apparatus, and moves over a disc of sensitive paper; this magnetic needle points, of course, to the north, and the luminous tube above it fixes its position on the sensitive paper, which is firmly fastened to the box. Therefore, it can be easily seen what situation the apparatus has assumed with reference to the magnetic needle—that is, the north.

Photographs of the Interior of the Eye, the Ear, etc.—Stein's
Heliopictor.

Medical science has already made great use of photography, both as a means of obtaining pictures of interesting anatomical preparations and phenomena of short
duration, and in giving exact anatomical views of the
different organs. The interior of living organs has been
disclosed by ophthalmoscopes, otoscopes, and laryngo-
scopes, and the image seen by the eye with these instruments
has been successfully photographed. Dr. Stein, of Frank-
fort-on-the-Maine, has done good service in this branch,
not only as a practical photographer, but also by the construction of suitable apparatus. It would exceed the
limits of this book to describe all the apparatus necessary
for this purpose. We shall content ourselves with the
description of one, that for taking the interior of the ear.

The apparatus (Fig. 83) consists of three parts : 1st,
the ear-funnel A ; 2nd, the illuminating apparatus B ;
3rd, the photographic apparatus D, with the lenses C.
These parts are placed together, as may be seen in the
accompanying diagram. The instrument is fastened by
a ball and socket joint to a suitable stand, in order to give
it the proper direction, according to the position of the
sun. The ear-funnel A is a conical tube about 1½ inches
in length, to push aside the small hairs which interrupt
the view; it is made of vulcanized india-rubber. The
illuminating apparatus B, which is easily closed by a
cover at a d, consists of two metal tubes, soldered to-

gether at a right angle at *b c*, of which one is provided
with parallel, the other with curved sides. At the place
where the two tubes unite is a perforated metallic mirror
(*e g f*), inclined at an angle of 45°.

The photographic apparatus consists of a double
objective *C* and a small camera, two inches deep. The
ground-glass screen *X*, and the dark slide *Y*, are fitted in
a frame *D*, easily moved. A plano-convex lens is placed
at *h*, between the objective and the illuminating apparatus. According to the position of the sun, bright
cloud, or any other source of light, the apparatus *B* can

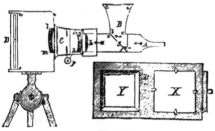

Fig. 83.

be moved by turning round on its axis; so that, in con-
junction with the joint of the stand, the apparatus can
be turned easily and steadily in all directions.

The rays which penetrate into the tube *B* are thrown
by the perforated plane mirror *e f* through *A* on the drum
of the ear. Reflected thence, they pass at *g* the perforated plane mirror, and the image of the drum of the
ear is thrown by the lenses on the ground-glass slide *n o*.
The image is focussed either by shifting the objective by
means of the screw at *p*, or by moving the lens at *h*, according as an enlarged image or one of life-size is desired.

During the photographic process, an assistant must pull the ear muscle backwards and upwards, in order to give a proper direction to the funnel in the tortuous aperture of the ear. The exposure in the sunlight, if a good collodion is employed, lasts half a second; under bright clouds on a clear day, from five to ten seconds, according to the intensity of the light. The exposure is effected by opening and closing a shutter at *c d*.

In order to render photography more accessible to physicians and scientific investigators, Dr. Stein has constructed an ingenious instrument called the "heliopictor," with which wet-plate photographs can be taken without any dark room. The heliopictor is a kind of dark slide which can be placed at the back of any camera. Dubroni, of Paris, first constructed such developing boxes. This box, a section of which is given in the diagram below, contains a glass vessel *K*, into which a silver solution can be poured through a stopcock, not visible in the figure. The glass plate to be prepared is coated with collodion, then brought through the door *T* into the box, and placed on the aperture *O* of the vessel *K*. The door *T* is then closed, and the plate is thus pressed by the spring *a* watertight against *K*. After this, the box is turned over to the right, the silver solution flows over the plate,

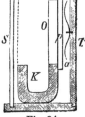

Fig 84.

and renders it sensitive. The course of this operation may be observed through a yellow glass slide *S*, which admits no chemical light. After the plate has been properly sensitized, the box is again placed upright, and brought into the camera in place of the ground-glass slide, S is drawn up, and thus the plate exposed. Then the silver solution is drawn off through a stopcock, and a solution of green vitriol poured in instead; by tilting the box this flows over the plate and develops the picture. The development may be watched through the yellow slide S: when it is completed, the picture is taken out and fixed.

Stein has improved his developing box by substituting a vulcanized india-rubber vessel, easily taken out and cleaned, for the glass receiver. He also introduced the method of filling and emptying the receiver by means of a stopcock, Dubroni having employed pipettes. Both apparatus are described in detail in the " Photographischen Mittheilungen," Jahrgang X. Nr. 117, 118.

SECTION VII.—PHOTOGRAPHY AND THE MICROSCOPE.

On Microscopes—Taking Microscopic Views—Their Application.

Nowhere has photography shown itself a more brilliant auxiliary to or substitute for the art of drawing than in the representation of microscopic objects. This was indeed attempted in the earliest days of photography, for Wedgwood and Davy attempted to fix the images of the solar microscope on sensitive silver paper. This solar microscope seemed, in fact, to be specially adapted for photographic purposes. Its construction is shown in Fig. 85. The microscopic object is inserted at m, and is either a drop of liquid upon a glass plate, or a small solid body compressed between two thin glass plates.

The small lens L projects an enlarged image of this

minute object on an opposite screen, or a white wall, exactly as shown in the following figure.

The rackwork at D serves to regulate the distance of the lens from the object m, and thereby to focus the object sharply upon the screen. E is a diaphragm by which the ragged edge of the round image is cut off. The tube $B\,C$ contains the lenses which concentrate the light. Each considerable enlargement diminishes the light of the picture materially; if it is enlarged three times, the light is diminished to $\frac{1}{9}$; if enlarged fourfold,

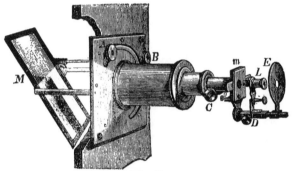

Fig. 85.

to $\frac{1}{16}$; if fivefold, to $\frac{1}{25}$; if a hundredfold, to $\frac{1}{10000}$. With such a diminution of brightness the eye would not see anything, if care were not given to throw an intense light on the object. The system of lenses contained in the tube $B\,C$ answers this purpose. This concentrates the sun's rays, which are reflected by the mirror M into the tube B, on the microscopic object; and the latter becomes in this manner so intensely illuminated that it admits of any amount of enlargement. The room in

which the instrument is placed is dark; accordingly, all
the conditions are present that enable proper photographs
to be taken.   All that is requisite is to expose a sensitive
plate in the place of the image

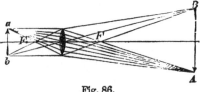

Fig. 86.

Few persons, however, possess a solar microscope.
For ordinary investigations a microscope is used similar
to that in the accompanying figure.   It contains at *o* a
system of lenses, which projects an enlarged image *S R*
(Fig 88) of the small object *r s*. This is
viewed through the eye-piece *c d* (Fig.
88), which is placed at *n* (Fig. 87), and
again magnifies the image *S R*, so that
a still greater image *S′ R′* (Fig. 88) is
produced.

This is seen directly by the eye of
the observer.   The necessary light is
thrown on the object by the help of a
concave mirror *s s′* (Fig. 87).

To produce photographs with the
help of such a microscope, a photographic camera, properly supported, can
be placed directly on the eye-piece *n*
(Fig. 87). This camera does not require
a lens like those previously described,

Fig. 87.

p. 88.   The eye-piece of the microscope serves this purpose, and is inserted through a light-tight sleeve into an opening in the camera. If, now, the tube h is raised, an enlarged image of the object at $S\,R$ will appear on the ground-glass screen of the camera, and can be easily photographed.

It is necessary in doing this that all light not emitted from the object should be excluded. If the mirror $s\,s'$ (Fig. 87) throws light on the object, many rays pass beside it, fall on the lenses, and occasion reflections that materially disturb the clearness of the image. In this case it is advantageous to insert a system of lenses between the mirror $s\,s'$ and the object, concentrating all the rays on the object.

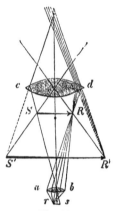

Fig. 83.

Instead of sunlight, an artificial light is sometimes used; for example, the electric or magnesium light, thus making the observer independent of the weather. The beauty of the microphotograph depends essentially on the beauty of the preparation to be photographed. This must be so arranged that it shows perfectly clearly all characteristic parts ; all disturbing accessories, dust and so on, must be removed, for they are equally magnified with the object. A skilful preparer is therefore necessary for successful micro-photography, which also depends on the excellence of the instrument, its proper arrangement, and the choice of the proper exposure. It is important in a microscope for this purpose that the lenses should be corrected for the chemically active rays. (See pp. 185, 186.)

Excellent results have been achieved in micro-photo-
graphy by Neyt at Ghent, Girard and Lackerbauer in
Paris, Fritzsch and Kellner at Berlin, and Woodward in
America.*

Microscopic photography is of the very greatest use
for preparing pictures of anatomical preparations which
change quickly, and of such chemical substances as
undergo rapid decomposition, also of the microscopic
crystals, which are enclosed in many kinds of stones, and
show themselves clearly in thin polished plates. In
the photographs the angles of these crystals can be
easily measured with the help of a protractor, and from
them the nature of the crystals may be inferred. Professor Gustavus Rose, in his treatise on meteorites, has
reproduced in steel engravings many micro-photographs
of this kind, taken by the author.

SECTION VIII.—MICROSCOPIC PHOTOGRAPHS AND THE PHOTOGRAPHIC
PIGEON POST.

Nature of Microscopic Photographs—Their Importance for Libraries--
Employment of the Pigeon Post.

Some years ago jewellery and toys were offered for
sale in Paris, containing small lenses in place of jewels.
If these were held before the eye, small transparent
pictures, some of them portraits, and others writings,
were visible. These little pictures were the so-called
microscopic photographs on glass. Such a picture is by
no means the representation of a microscopic object, but
of a large-sized object, only it is so small that a micro-

* Full details are given in "Die Photographie, als Hülfsmittel mikro-
skopischer Forschung," by Moitessier and Bennecke. Brunswick:
Vieweg.

scope is required to see it. The production of these
photographs does not differ from that of others; it only
requires an instrument forming images of microscopic
minuteness, and this is effected by employing small
lenses of very short focal length. In using these a direct
photograph is not taken, but in the first place a photographic negative is prepared with an ordinary camera
from the object chosen; after this, with the help of the
small lens, microscopic positives on glass are obtained
with the ordinary collodion process. These are then cut,
and a small lens fastened on them, and then they are
mounted in metal. Such pictures are in themselves
little else than toys, which have, however, been put to a
bad purpose, for indecent photographs have been thus
placed in the hands of unsuspecting persons—a fact that
speedily brought this branch of photography into dis-
credit. But there are circumstances in which such
microscopic photographs can be of extraordinary value.
Simpson in England has called attention to the fact that,
by the help of photography, the contents of whole folios
can be concentrated within a few square inches, and that
the substance of books filling entire halls, when reduced
by microscopic photography, can be brought within the
compass of a single drawer—a circumstance which, with
the enormous increase of material that has to be stored
by our libraries, may be of importance. Of course, either
a microscope or a magic lantern would be required to
read such microscopic works.

Hitherto photography has not been applied to this
purpose, though Scamoni's heliographic process, described
further on, would considerably facilitate the creation
of such microscopic libraries. Microscopic photographs,

however, have obtained great importance in promoting pigeon despatches. During the siege of Paris in 1870, the blockaded city held communication with the world outside by means of balloons and carrier pigeons. The first mode of communication was used almost exclusively for political objects; the second only admitted the transmission of very light letters. Letters, however condensed, could scarcely have been sent more than two or three at a time by a pigeon. In this case, microscopic photography presented a valuable means of concentrating many pages on a collodion film of only one square inch, and so light that more than a dozen packed in a quill could be sent by one pigeon. Dagron, at Paris, who first prepared microscopic photographs, also was the first to prepare these pigeon despatches. All the correspondence which had to be diminished was first set up in type, and printed together on a folio page. A microscopic photograph was made of this folio page, contained in about the space of $1\frac{1}{2}$ square inches. The collodion film of the picture was then removed from the glass by pouring a collodion over it which contained castor oil in solution. The collodion soon dries, and can then be separated from the glass, carrying the film of the picture with it. Such films contained in some cases as many as 1500 despatches. At the place of arrival the films were unrolled and then enlarged by the help of a magic lantern; a number of writers thereupon set to work to copy the enlarged despatches, and ultimately forwarded them to their respective addresses. Thus Paris corresponded, by the aid of photography, for six months with the outer world, and even poor persons were able to let their relatives know that they still lived.

SECTION IX.—PYRO-PHOTOGRAPHY.

Pictures on Porcelain—Their Production by Photography—Grüne's Method—Its Application for the Decoration of Glass and Porcelain.

An ordinary photograph is, as paper, very combustible, and exposed to injury from corrosive substances. Encaustic images on porcelain and glass do not participate in this exposure to injury, and therefore attempts have been made to prepare fireproof photographs, especially for the decoration of glass and porcelain. Success has crowned these efforts in several cases. One of the simplest processes is that of W. Grüne at Berlin.

Grüne found that the collodion image—which, as we have seen (p. 109), consists of minute parts of silver—is capable of manifold changes, and that, moreover, it is easily transferable, by means of the elastic collodion film, to other bodies. The film, with the picture, can be removed from the glass, placed in different solutions, and then transferred to curved surfaces, etc. If the collodion picture is placed in a metallic solution, a chemical change ensues. Assuming the solution to contain chloride of gold, then the chlorine passes over to the silver, of which the picture consists, chloride of silver is formed, and metallic gold is precipitated as a fine blue powder on the picture. Thus a gold picture is obtained.

With certain precautions this can be transferred to and burnt into porcelain. By this means a dull image is obtained, which can be rendered brilliant by polishing. Grüne has employed this method to produce gold ornaments on glass and porcelain. Drawings and patterns of various kinds are photographed; the image obtained is changed into one of gold, then burnt in, and thus the most beautiful and complicated decorations can be produced without the assistance of the porcelain painter.

If a silver picture be plunged into a solution of platinum instead of a solution of gold, a platinum image is obtained. This assumes a black colour on being burnt into the porcelain. In this manner black portraits, landscapes, etc., have been produced on porcelain.

Pictures of this kind can be prepared in other colours than black. For example, if the photograph is dipped in a combined solution of gold and platinum, the gold and platinum are precipitated on the picture. The image thus obtained, if burnt in, presents a very agreeable violet tint.

Solutions of uranium, of iron, and of manganese effect precipitates on a collodion picture, modifying its colour, and, when burnt in, produce different brownish or blackish tints. We shall see, later on, that there are other means of producing such pyro-photographs. Details will be found in the chapter on the photo-chemistry of the chromium compounds.

SECTION X.—MAGIC PHOTOGRAPHY.

Invisible Photographs—Their Development—Magic Pictures and Magic
Cigar Holders.

Closely connected with Grüne's process for producing pictures on porcelain, is what is called magic photography. A few years ago small sheets of white paper were offered for sale which, on being covered with blotting-paper and sprinkled with water, displayed an image as if by magic. The white sheets of paper, to all appearance a blank, were photographs which had been bleached by immersion in a solution of chloride of mercury. If a photograph not containing gold—all the usual paper photographs contain gold—be immersed in a solution of chloride of mercury, a part of the chlorine passes over to the silver of the picture, and changes the brown silver particles into white chloride of silver, which is invisible on the white paper.   At the same time subchloride of mercury (mercurous chloride), which contains less chlorine than the chloride of mercury, is precipitated. This body is also white, and therefore invisible on the white paper. Now, there are several substances which colour this white subchloride of mercury black.   Among these are ammonia and hypo-sulphite of soda.   If, therefore, the invisible picture is moistened with one of these substances, it is coloured black and becomes visible.   In the magic photographs formerly sold there was hypo-sulphite of sodium in the blotting-paper ; this was dissolved on moistening the paper, the solution penetrated to the picture and made it visible.

Quite a different kind of magic photograph was offered for sale some years later—the magic cigar-holders. These contained a small sheet of paper between the cigar and the mouthpiece, which was exposed to the action of the cigar smoke ; with continued smoking an image became visible on the sheet of paper, which contained a magic photograph of the kind described above.   The image was brought out by the vapour of ammonia which is contained in the smoke, and which has also the property of colouring the magic photographs black.

The magic photographs of recent times were introduced at Berlin by Grüne, but their principle was known before, as J. Herschel had produced similar ones in 1840.

SECTION XI.—SCAMONI'S HELIOGRAPHIC PROCESS.

Defects of the Silver Positive Process—Relief of the Photographic
Negative—Impress of the same on Copper.

It was stated at an earlier page, that the time
required to produce photographic positives formed a
serious defect in the process. Every picture that has to
be printed from a negative must have a longer or shorter
exposure to the light, and the worse the light, the longer
the time required. This is of no moment with a dozen
portraits, but if hundreds or thousands are to be prepared,
time is of consequence.

Another disadvantage in the silver print is its high
price and doubtful durability. Attempts have been made,
since the discovery of photography, to overcome these
defects by combining it with printing-press processes—
lithography or metal-plate printing. The ordinary process of printing from a metal plate is as follows. The
drawing is engraved upon a copper or steel plate; this
is covered with engraver's ink, the ink penetrates into
the incisions, and under a heavy press passes over to the
paper, thus forming the engraving. Impressions of this
kind can be made in a short time in great quantities,
without the help of light, and without employing expensive salts of silver. We showed in the first chapter
(p. 10) that an incised drawing on a metal plate can be
made with the help of photography by coating it with
asphalt. But the same object can be attained in other
ways; and one of the most original is that of Herr G.
Scamoni, at St. Petersburg, the able heliographer of the
Russian Imperial Office, for the preparation of State
papers.

He observed that an ordinary photographic negative does not form a plane surface, but appears in relief; the transparent places—shadows—being deeper than the opaque. But the difference is very slight. Scamoni tried to increase it by treating the freshly taken and developed picture with pyrogallic acid and solution of silver. In this manner fresh silver was precipitated on the picture, which has the property of attracting and retaining silver separated chemically. The relief was considerably incased by this strengthening process. It can be augmented by a treatment with solutions of chloride of mercury and iodide of potassium, which transform the metallic silver into more bulky compounds. A relief was ultimately obtained nearly as high as the incisions of an engraved plate are deep. If, now, a collodion positive be taken of a negative from a linear drawing, and treated as above described, we have all the requisites for producing from it an engraved copper plate. The relief-like photographic plate is brought into a galvano-plastic or electrotype apparatus, of which we shall speak further on. This produces on the plate a coherent copper precipitate, which is in low relief where the plate shows high relief; that is, where there are strokes or outlines. Thus a copper plate is obtained from which impressions can be taken as well as from one that has been engraved. This process is now used to reproduce drawings like copper plate.

Maps are prepared in this manner, in which the original can be photographically enlarged or diminished; also writings on an enlarged and diminished scale. Scamoni has thus reduced a page of the illustrated journal, " Ueber Land und Meer," to sheets of one inch in width, on which the print is perfectly legible through the microscope. Feats of this kind are not mere play, but they have a great importance for the preparation of paper money and for libraries, as we showed at pp. 11 and 205.

SECTION XII.—PHOTOGRAPHY AND JURISPRUDENCE.

Photography as a Means of Identification—Photographs of Criminals, of Railway Accidents, Fires, Documents, etc.

The application of photography to jurisprudence is of great interest. The faithful likeness of a man, or of an object, makes their recognition more certain than the most circumstantial description in words; and photography gives us such likenesses. Accordingly, repeated attempts have been made to utilize it as a means of identification. This was first attempted in 1865, when the season tickets for the photographic exhibition at Berlin contained the portrait of the holder, that they might not be transferred. This plan is now adopted in the season tickets of the Zoological Gardens at Berlin. It is still more important for the recognition of criminals. Old offenders are now photographed in prisons, partly as a means of recapture in case of escape, partly to detect them in case they should be again brought in under a false name.

Justizrath Odebrecht, in a treatise on jurisprudence, recommends that photographs should be taken of bodies found dead or, in case of murder, of the victim and the surroundings, for the information of judges. This has been repeatedly done. Further, the scene of a railway accident, buildings that have been destroyed by fire or storms, are photographed for the information of railway and insurance companies, or of the legal authorities. Photography is of especial use for this purpose since the pictures may be completed within a few minutes, and made even during the repairs. It is also of value in jurisprudence for detecting forgeries. Very frequently forged cheques are photographed in order to send a copy for the information of those interested. Stolen and recovered articles are also often photographed to bring them to the notice of the proprietor. In many large cities the police cause pickpockets and sharpers to be photographed, and show an album of this kind to persons who have been robbed.

SECTION XIII.—PHOTOGRAPHY, INDUSTRY, AND ART.

Photography as a Means of Artistic Culture—Extension of the Art of Drawing through Photography—Pattern Cards—Building Plans.

We have already laid stress upon the importance of photography as a means of reproducing works of art. It makes every work of art accessible to persons of slender means, and therefore it has become as important an auxiliary for popular culture in the province of art as the printing-press is for science.

Photography is equally important in those branches of industry in which accurate plans are indispensable; for example, architecture and the construction of machinery. In their case photography forms an addition to the art of drawing, effecting in a few minutes what the draughtsman could only accomplish in several hours or days, and copying with a faithfulness to which no draughtsman could attain. In connection with this subject we have already described, in our third chapter, the technical importance of the *lichtpaus*, or new Talbot-type process. It is the easiest kind of photography, but it only gives copies of the size of the original; however, the negative process allows an enlarged or diminished copy to be taken of every drawing, according to option. Photography is already very generally used for these reproductions. It is equally important for taking views direct from nature, be they machines or parts of machines, buildings or parts of buildings. Pictures of this kind present not only a graphic image, but they serve for instruction and demonstration in lectures. The photograph of a house, of which the length, depth, and height are known, supplies sufficient data for the dimension of any part of it if the perspective be taken into consideration. Pictures on a small scale are commonly sent out as specimen cards. Iron foundries, manufactories of bronzes and of porcelain frequently issue price lists with photographic illustrations, which are multiplied from negatives of originals by the Albert-type process (See the following chapter.)

Further, an original application of photography has been made by architects who are at a distance from a building under their direction. They cause photographs to be taken every week, giving them a clear picture of the progress of the building. We have already hinted at the services that photography can render in the manufacture of porcelain, and further in combination with the multiplying arts. We shall learn more on this subject in the following chapter.

CHAPTER XV.

PHOTOGRAPHY WITH CHROMIUM COMPOUNDS.

WE have given a full account, in the first part of our book, of the chemical and physical principles of photography with salts of silver, and of its application to art, science, life, and industry.

Numerous attempts have been made to substitute other sensitive materials for the expensive salts of silver, and some of these attempts have been crowned with success. It is true that no substance has been hitherto found permitting a negative to be prepared in the camera as easily as iodide of silver. For the production of camera pictures from nature, we have no substitutes for iodide and bromide of silver. But positives from negatives already existing can be successfully produced, not only by the help of salts of silver, but also of other metallic combinations. The results obtained are indeed inferior in beauty to the silver pictures, but we shall see, later on, that they admit of multiplication in the printing-press without the help of light. We shall now describe the most important of these processes.

Oxides (Combinations of Chromium with Oxygen)—Salts of Chromium
—Chromic Acid—Chromic Acid Salts in the Light—Ponton's
Discoveries.

A black mineral called chrome iron ore occurs in
nature, especially in Sweden and America. If this is
fused with carbonate and nitrate of potash, a beautiful
orange-red salt is formed, which dissolves in water and
readily crystallizes on evaporation. This orange-red salt
is bichromate of potash. It consists, as implied by the
name, of chromic acid and potash. The latter is the
chief component part of the soap-boiler's ley; the former
is a combination of a metal, like iron, with oxygen.
Chromium and oxygen combine together in several
proportions, thus :

28 parts chromium with 8 parts oxygen form protoxide of chromium.
28   ,,   ,,   ,, 12 ,,   ,,   ,, sesquioxide of chromium.
28   ,,   ,,   ,, 16 ,,   ,,   ,, dioxide of chromium.
28   ,,   ,,   ,, 24 ,,   ,,   ,, chromic acid.

The last combination, chromic acid, is the best known
of all. It may be prepared by adding sulphuric acid to
a solution of chromate of potash : it crystallizes in red
needles, which readily part with their oxygen. Thus
alcohol, dropped upon chromic acid, takes fire; the
chromic acid instantaneously gives up part of its oxygen
to the alcohol, and is changed into a green substance,
sesquioxide of chromium. The sesquioxide of chromium
forms salts with acids; for example, sulphate of chromium.
This unites again readily with sulphate of potash to form
a double salt, which is known by the name of chrome
alum, and is sold crystallized in very beautiful dark

violet octahedra.   It is employed in dyeing, together with chromate of potash.

If chromate of potash be mixed with a solution of protosulphate of iron (green vitriol), a brown precipitate of peroxide of chromium is formed, the iron salt having removed a part of the oxygen from the chromate.   This is often formed by the action of substances absorbing oxygen on chromic acid or its salts.

Chromic acid is of special interest to us, because both it and its salts are sensitive to light.   Neither pure chromic acid, nor chromate of potash, are changed by light; they can be exposed for years to the sunlight without any decomposition being perceived.   But as soon as a body is present that can unite with oxygen—for example, wood-fibre, paper, etc.—the light immediately produces its effect. This fact was published in the same year as the discovery of photography, in 1839, by Mungo Ponton, in the " New Philosophical Journal."   He writes :—

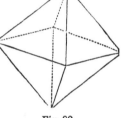

Fig. 89.

" Paper immersed in bichromate of potash is powerfully and rapidly acted upon by the sun's rays. . . . . . When an object is laid in the usual way on this paper, the portion exposed to the light speedily becomes tawny, passing more or less into a deep orange, according to the strength of the solution and the intensity of the light.   The portion covered by the object retains the original bright yellow tint which it had before exposure, and the object is thus represented yellow upon an orange ground, there being several gradations of shade or tint according to the greater or less degree of transparency in the different parts of the object.

"In this state, of course, the drawing, though very beautiful, is evanescent. To fix it, all that is required is careful immersion in water, when it will be found that those portions of the salt which have not been acted on by the light are readily dissolved out, while those which have been exposed to the light are completely fixed on the paper. By this second process the object is obtained white upon an orange ground, and quite permanent. If exposed for many hours together to strong sunshine, the colour of the ground is apt to lose in depth, but not more so than most other colouring matters."

It appears that Ponton made experiments similar to those of Talbot, in the first period of silver photography. Perhaps he also copied leaves (see p. 5). The copies which are produced in this manner on chromate of potash are, however, immeasurably fainter than the copies on silvered paper.

They may be easily prepared by exposing under dried leaves, a drawing, or a negative in a printing frame, pieces of paper which have been soaked in a solution of chromate of potash for about a minute, and then hung up to dry in the dark or by lamplight.

The chromic acid is then reduced to brown peroxide of chromium, but if the exposure lasts very long the reducing process goes further, and a green oxide of chromium is formed. In this case the picture appears fainter.

Ponton's experiment remained a mere curiosity until the inventor of photography on paper containing silver salts discovered another property of chromate of potash, which led to the most extensive applications.

This property consists in the action of the combinations of chromium on glue.

Glue in its purest form, known by the name of gelatine, is insoluble in cold water, but it absorbs cold water like a sponge, and thereby swells. If it is warmed with water, it dissolves; but on cooling the solution solidifies to a jelly. This property is used to thicken soups. If to the warmed solution of glue be added alum, or a salt of the oxide of chromium, or chrome alum, the glue becomes insoluble in water, and forms a precipitate. On this is based the well-known system of white tanning; for in the tanning of a piece of leather the alum combines with the gelatine contained in the leather, and this becomes thereby insoluble, and at the same time durable.

Bichromate of potash and glue can be dissolved together in warm water in the dark, without the glue suffering any change from the chromate. If a plate or a sheet of paper be covered with such a solution of gelatine and bichromate of potash, and the film be allowed to dry, it becomes firm, and yet remains soluble in water as long as it is kept in the dark. But as soon as the film is exposed to the light, the bichromate of potash is reduced to oxide of chromium, and this tans the film of gelatine; that is, makes it insoluble in water.

This observation was made by Fox Talbot in 1852, and, as a careful observer, he knew directly how to turn it to account. He coated a steel plate with a solution of chromate and gelatine, let it dry in the dark, and then exposed it under a drawing or a positive glass picture. The black lines kept back the light. Accordingly, at these places the gelatine remained soluble, but was rendered insoluble under the white places, by the action of light. After exposure he washed the plates in the dark with warm water. By this means the places that had remained soluble under the black lines were dissolved; the others were retained on the plate. Thus Talbot obtained a drawing on the metal itself on a brown ground. This is worthless by itself, but it provides the means of producing a steel plate for engraving.

We have already explained, at p. 210, the nature of steel and copper-plate engraving. Both processes consist in the production of a metal plate which contains, in incised lines, the drawing that is to be reproduced. These lines retain the ink which is rubbed upon the plate, and transfer it to the paper. The hard steel plates have the advantage of lasting for many more copies than the softer copper plate; only the steel engravings are far inferior to copper-plate in artistic beauty, and therefore the former have lost favour. But the steel engraving is very important for preparing technical and scientific diagrams, paper money, and the like, as less artistic beauty is required in their case. It was steel plates of this kind that Talbot produced by the help of light.

We have seen that his steel plate was covered with

an insoluble film of gelatine, and that the metal was uncovered at all places where the light had not operated. He poured on it a fluid which ate into the steel; for example, a mixture of acetic acid and nitric acid. This mixture, of course, only attacked the steel where it was exposed, and thus produced an incised drawing in the steel plate, so that the latter, after being cleaned, gives as good an engraving as if it were the work of the engraver.

Thus a new process was found to replace the difficult work of the engraver by the chemical action of light.

We have mentioned in the first chapter a similar process, based on the application of asphalt and a different one by Scamoni. (See p. 210.)

This discovery of Talbot was soon followed by a more productive one on the same ground.

An Austrian, Paul Pretsch, prepared, in 1854, copper-engravings by a similar method, with the help of the electrotype process. He also took a film of gelatine, which contained chromate of potash, exposed this under a negative or a positive picture, and then washed it in hot water.

After doing this, all the places were retained which had become insoluble through the light, and after the washing and drying they stood out in high relief.

Accordingly, in copying under a positive, the lines which were black in the original appeared in low relief, and the white parts in high relief.

This film was placed in an electrotype apparatus. This apparatus has the property of precipitating copper or other metals on a surface. It consists of a galvanic cell, as described p. 70, and a trough containing a solution of sulphate of copper. The reliefs which are to be copied, having been made conductors of electricity by a coating of graphite, are suspended from the zinc pole *B* (Fig. 90) of the battery in the sulphate of copper: to the copper pole *D*, a plate of copper is attached. As soon as the galvanic stream passes, the solution is decomposed, and the copper adheres to the relief. The thickness of the copper depends on the time the current is allowed to last; accordingly, plates of any thickness can be produced.

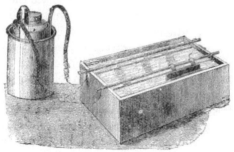

Fig. 90.

If the original was in low relief, the electrotype impression will be in high relief, and *vice versâ*. Therefore, in the above case an impression is obtained with lines in high relief.

This kind of plate is also adapted to give impressions, but rather differently from an incised copper plate. In the latter, the engraver's ink is rubbed into the incised marks, and then under strong pressure conveyed to paper.

In a plate with a drawing in high relief, the impression takes place as in printing; the raised places are rubbed over with printer's ink by the help of a leather ball, or of a cylinder blackened with ink, and then printed on paper. Letter-press is produced in this manner; all its letters are in high relief; also all woodcuts which accompany the text.

The printing-press is the simplest and cheapest mode of multiplying copies. It admits of the use of cheap papers, whilst copper-plate engraving requires a thick, soft, special paper. The printing-press, moreover, admits of woodcuts being printed in the text, whilst copper-plate printing requires special plates. Lastly, the printing-press works with extreme rapidity, whereas copper-plate printing requires much more time.

Further, the printing-press does not use up the type rapidly, as it works under feeble pressure; while copper-plate printing, which requires strong pressure, wears the plate considerably, so that after striking off a thousand copies, the impressions are no longer as good as at first.

The production of photographic plates for the printing-press is of the greatest importance, and Pretsch has taken the lead here. His process did not, indeed, produce the most perfect results. The low-relief plate which he produced on the gelatine film by the help of light was not deep enough to produce a high relief in the galvanic impression; but this is necessary, for otherwise the printer's ·ink penetrates into the incised parts, which ought to remain white, while the washing of the exposed gelatine films with hot water easily destroys the finer parts of the picture, and this detracts materially from the value of the copies. Moreover, difficulties arise in preparing the electrotype plates; the film swells up and loses its form. In short, the affair is not so simple as it appears; little

difficulties exist, and these occasion errors which the
unprofessional hardly observe, but which considerably
diminish the effectiveness of the picture.

At an early period it was found that these processes
offered a special difficulty, viz., the reproduction of the
transitions from light to shade—the half-tones.   These
were so very imperfect that the representation of natural
objects—portraits and landscapes—was speedily given
up, and people confined themselves to the reproduction of
drawings, maps, and the like, on an enlarged or diminished scale, and thereby to producing stereotype plates
for copper engraving and printing.   This application is
of no little importance, for it prepares a metal plate for
printing, by the help of light, in as many hours as an
engraver requires days, and at far less cost.

We add two plates to the present work, which have
been prepared by Scamoni of St. Petersburg, by the help
of gelatine and bichromate of potash, by a modification
of the process just described.   Both are impressions from
heliographic plates: the smaller one—Plate III., "Am
Rhein"—is a plate in high relief, which is printed in the
printing-press; the other—Plate IV., "Johannisfest"—a
low relief, in the style of copper engraving.

SECTION III.—THE PRODUCTION OF PHOTO-RELIEFS.

Photo-sculptures—The Pantograph—The Fount Process—Chromogela-
tine-relief—Fount-relief by Cold Water—Relief by Cold Water—
Difficulty of its Production—The Transfer Process.

More than ten years ago intelligence was received
from Paris of an entirely new discovery—photo-sculpture
—which was said to produce statues by the help of light.

According to the description, this was effected as

PLATE III.

P. 224

ELECTROTYPE FROM HELIOGRAPH.

Plate IV.

COPPER PLATE FROM HELIOGRAPH.

follows: a person was placed in the middle of a circle of about twenty photographic apparatus, which at a given moment took twenty pictures of the person, and represented him on every side.   These photographs were afterwards transferred to clay, by means of an instrument commonly called a pantograph. This consists of a system of bars *a b c d* (Fig. 91).   Of this system one bar can be turned about a pivot *x*, the others are moveable at the joints; *m n* are two pencils.   If one pencil *m* is carried

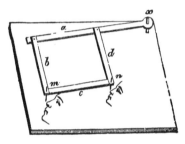

Fig. 91.

along a drawing, the other pencil *n* makes the same movement, and, if a piece of paper be placed under it, *n* draws exactly the same line which the first pencil *m* describes.   If instead of the pencil *n* we conceive a knife, which cuts out in clay the outline described by *m*, a profile is obtained in clay by moving *n* along the outline of an image, and in this manner all the outlines of the person taken can be transferred to clay.   This photo-sculpture, as it is called, can only be carried out imperfectly.   A careful manipulation by a very clever artistic hand is necessary for the work, and is indeed the essential point.   As far as the author has examined the

Q

matter, the pantograph is a mere ˋpretence. A clever artist models the bust according to the photographs.

Nevertheless, these reliefs can be produced by light, and these reliefs are not inventions of advertisers; they are easily produced, and it is surprising that the process has not yet made a stand.

We have mentioned above the properties of gelatine, and remarked that it has the power of absorbing cold water and thus swelling up. This property is lost if the gelatine is saturated with chromate of potash and exposed to the light. If this exposure is made under a negative, all the places situated under the transparent parts lose this property, while the other places not affected by the light retain it. Accordingly, if the exposed film be thrown into water, the places which are not affected by light swell, whilst those affected by light are not altered. The result is a true relief,—the lights are in high relief, the shadows in low relief,—and this is so strong that it can be cast in plaster of Paris. For

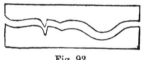

Fig. 92.

this purpose the relief is dried with blotting-paper, rubbed with oil, and then a paste of plaster is poured over it. This soon hardens, and gives an impression of the gelatine relief, being in high relief where the gelatine is in low, and *vice versâ* (Fig. 92).

It appears as if a plate might be easily obtained for letter-press from such a gelatine relief. Let the gelatine film under a drawing be supposed to be exposed. The black lines then keep back the light; accordingly, the gelatine particles come out in high relief on being wetted with water. The drawing is therefore represented in high relief, and this is exactly what the printer requires; nothing further would now be required than to cast the relief in plaster, and recast the plaster form in metal, as happens daily in the stereotyping of woodcuts. But unfortunately this process breaks down, owing to a trifling circumstance—the lines in the wet relief do not swell to the same height. But the printing-press requires them to be perfectly level, otherwise they cannot be equally inked and printed.

On the other hand, the casting can be very well utilized as a picture in relief, if suitable retouches are given to it. Reliefs of this kind with portraits were sold some years ago for seals, but the execution was very imperfect, and therefore they soon lost favour.

But reliefs can be obtained in another way from an exposed gelatine film, namely, by hot water. As we have seen above, this dissolves the parts which, not having been affected by light, have remained soluble, while the parts affected by light, and therefore insoluble, remain untouched. These parts that have remained insoluble stand out as prominences.

Another precaution is necessary. Suppose that N (Fig 93 a) is a negative, that c c are its transparent parts, and b the semi-transparent, what are called half-tones. If a film of prepared gelatine g (Fig. 93 b) is exposed under them, the light penetrates in various degrees, according to its strength—most in the transparent places, less in the semi-transparent, and not at all in the opaque parts.

Accordingly, insoluble films of different thickness will be formed, as represented in b. The shaded parts in the figure denote the portions that have become insoluble.

If now the film of gelatine (Fig. 93 b) is plunged in

hot water, all the parts left white in the figure become
dissolved; but at the same time the half-tones not adhering to the substratum P—for instance, paper—become
detached and are torn off. Therefore a relief of the form
d remains behind; the half-tones ($y\ y$) are wanting. In
order to avoid this a support must be given to the
exposed surface to retain the half-tones. For this pur-

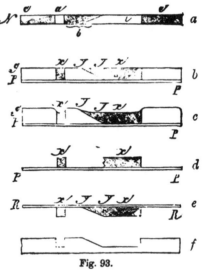

Fig. 93.

pose a piece of paper, moistened with white of egg, is
laid on the exposed surface and becomes firmly attached
to it. If the sheet is now immersed in hot water, the
film P becomes detached from g, the little portions
of gelatine remain attached to the albuminized paper;
the white places in Fig. 93 b become dissolved, and all
the half-tones $y\ y$ adhere firmly to the new layer, as

in Fig. 93 *e*, and form a relief. This is named the transfer process. If the relief produced by cold water, described p. 226 (Fig. 93 *e*), is compared with that produced with hot water (Fig 93 *c*), the difference is at once apparent: in the former case the parts not affected stand out in relief, in the latter case those exposed to light.

SECTION IV.—PRINTING IN RELIEF.

Production of Photographic Half-tones—Production of a Printing Plate in Relief from a Gelatine Relief—Woodbury's Printing Process—Its Importance—Printing in Relief on Glass—Magic-lantern Pictures.

Although production of reliefs with cold and also with hot water, described in the previous chapter, has not at present been utilized for any kind of photo-sculpture, a new printing process has been founded on it, which was invented by Woodbury in 1865, and has been called Woodbury-type, from its inventor.

The heliographic methods of printing previously described are apparently very simple. Pictures of all objects cannot, however, be prepared by these methods. A linear drawing, such as a map or letterpress, can be reproduced either on an enlarged or diminished scale, by these methods; but pictures with half-tones, such as sepia drawings or photographs from nature, cannot be so copied. The soft half-tones become rough and hard, rendering the picture very ugly. According to Osborne, the cause of this is found chiefly in the nature of half-tones in copper-plate printing, which are formed by a series of lines at various distances from each other or by roughening the plate; the latter method produces a series of points, which, according to their distances from each other, give a lighter or darker shade or half-tone. The half-tone of sepia drawings and photographs is quite different. It is not formed of strokes or points, but is a homogeneous light or dark colour.

Accordingly, it was first necessary to break up the photographic half-tone into a series of strokes or points, to make it a copper-plate half-tone, and this constitutes the difficulty.

Woodbury conceived the idea of producing, by a new printing process, homogeneous half-tones, perfectly similar to those of photographs or sepia drawings.

He produced a relief by exposing a film of prepared gelatine under a negative, and then treating it with hot water. (See last chapter.) This relief shows the dark parts of the original in high relief, and the light parts in low relief. For the negative is transparent where the original is black; hence the light passes unimpeded through those places. The half-tones are formed by the varying height of the gelatine film. (Compare Fig. 93 *e.*)

If this gelatine relief is suffered to dry, it becomes wonderfully hard and firm. It can then be placed with a plate of lead under a strong press, and a *cliché* of the relief can be thus obtained in lead. The prominent parts of the gelatine relief appear, of course, depressed in the lead, and the depressions prominent, as represented in Fig. 93 *d.*

Woodbury uses this lead relief as a printing plate. But he does not print it off with oily printer's ink, which is too opaque, but with a semi-transparent gelatine ink. This is poured warm on the plate in a horizontal position, it penetrates into the depressions, and now, if a piece of paper be placed upon it and pressed gently down, the gelatine consolidates quickly, and an impression in relief is obtained on the paper. As the ink is transparent, it appears in thin sheets much less black than in the thick, and in places where its thickness gradually diminishes occurs a transition from black to white—a perfectly homogeneous half-tone. As soon as the coating dries, the relief contracts considerably, but the semi-transparency remains, and thus it is possible to reproduce the most beautiful half-tones of photography by printing. Any colour may be used in this process.

This relief-printing of Woodbury has already attained a high importance. It makes the multiplication of photographic negatives from a single printing form possible without the help of light. It is therefore of importance where a great number of pictures are required; for example, in the production of copies of oil-paintings and drawings. Photographers do not use it much in portraiture, because the production of a faultless gelatine relief and its impression on lead require great practice and an expensive apparatus, which would not pay in the limited sphere of portrait photography.

The frontispiece of the present work, representing the moon, from a photograph by Rutherford (mentioned p. 188), is an impression in relief, by the Relief Printing Company in London.

It is a special advantage of the relief-printing process that it admits of printing on glass. Wonderful transparencies are thus obtained, very effective as window ornaments. Goupil has prepared copies of oil-paintings by this method, and they are frequently to be seen in the windows of our dealers. The transparent stereoscopic pictures on glass, produced by this process, are

equally charming—in sharpness and softness they almost
excel the ordinary silver copies.   Recently a number
of beautiful magic-lantern pictures, produced by the
Woodbury process, have been offered for sale, and will be
eventually used as an important means of instruction in
schools.   The author has a collection of American landscapes of this kind, which, when enlarged in the magic
lantern, are more instructive than the fullest geographical treatise.

Pictures of this kind can be sold much cheaper than
the ordinary transparent photographs for stereoscopes.
(See the chapter on landscape photography, p. 155.)

SECTION V.—PIGMENT PRINTING.

Poitevin's Process—Production of Pictures in any Kind of Pigment—Its
   Difficulty—Inverted Impressions—Transfer Process—Comparison of
   Pigment and Silver Prints—Braun's Facsimile of Coloured Draw-
   ings—Transfer by Contact.

WE have seen above that gelatine mixed with bichromate
of potash becomes insoluble in the light.   This fact was
made by its discoverer, Talbot, the basis of heliography—
that is, of photographic steel engraving.   Poitevin, a
Frenchman who has done much to promote photography,
founded on the same method another process : he produced pictures in various colours.   He first used carbon
as a pigment, thus obtaining carbon pictures.

The process is simple : Poitevin coated paper with
prepared gelatine coloured with lampblack, and exposed
this under a negative ; he then washed the film in hot
water, which dissolved those parts of the gelatine un-

affected by the light, whilst the insoluble parts retained their colouring matter and thus formed a picture.

Practical difficulties occur even in this simple process. As has been already mentioned, the action of the light does not always penetrate the whole thickness of the film. The half-tones have, therefore, no support, and are torn off in washing (see Fig. 93 *e*, p. 228). Before treating the films with hot water, it is therefore necessary to transfer them, as described in the chapter on photo-reliefs. An albumenized sheet is pressed in the dark on the coloured film of gelatine, and then the whole is plunged into hot water; the half-tones adhere to the paper pressed upon them, and the image appears uninjured on it, as in Fig. 93 *e*.

The picture is, of course, reversed; that is, what was originally to the right in the lower image comes now to the left. That this is so can be easily seen by writing a word with thick ink in large letters on a piece of paper, and laying a piece of blotting-paper on the wet writing. The reversed impression of the writing is seen on removing the blotting-paper. In the letter copying-press the same thing takes place, and therefore the letters are copied on very thin paper, that they may be read on the reversed side, because viewed from that side they appear in their first position. The gelatine prints cannot be printed on such thin paper; therefore, if the reversed position is inconvenient, the picture must be again transferred. This has been latterly managed in the following manner :—

The moist gelatine film is placed after exposure upon a smooth zinc plate, to which on drying it becomes very firmly attached. The copy thus glued to zinc is immersed in warm water, the paper becomes detached, and the developed image adheres to the zinc plate. A sheet of white paper is now fastened with glue upon the zinc plate, and allowed to dry. The gelatine picture adheres firmly to this paper, and may, with care, be detached from the zinc plate. The picture thus appears unreversed upon the paper. This more recent form of the process is employed in England at Woolwich Arsenal.

The pictures thus obtained are very similar to those of the Woodbury-type; they surpass the latter, however, in the fineness of their details and the ease with which they may be produced.

This process has not yet supplanted silver printing, for the expense of the material, owing to the twofold use of paper, equals that of silver photography, and the labour, being somewhat more complicated, is therefore dearer. The pigment impressions have a great advantage in the fact that they can be produced in any colour; genuine Indian ink may be used for them, and then perfectly durable pictures are obtained that do not turn yellow or black.

In the same way red, sepia, blue, and so on, can be mixed with the gelatine, and thus pictures can be produced in those colours. This circumstance is important when it is wished to reproduce coloured drawings. Quantities of such drawings and sketches of the old masters are in various museums. Braun of Dornach, in Alsace, the same photographer who made himself conspicuous for his Swiss views, has undertaken to reproduce such drawings in their original colour by the process just described, preparing first a silver negative in the usual way, and printing from this on coloured gelatine films.

He has thus made cheap facsimiles of many drawings of which no copy had previously existed.

Latterly, a very interesting observation has been made by Abney, in England, in relation to the foregoing process. He remarked that if an exposed film of gelatine remained a long time in the dark the insolubility increases. Accordingly, a film of this kind which, freshly developed, would only give a faint image, after some hours gives a strongly defined image. This fact allows the time of exposure for pigment pictures to be considerably reduced; that is, a larger number of pictures to be made in the same time.

Still more interesting is an observation of Marion, at Paris. He exposed under a negative a sheet of paper sensitized with bichromate of potash, and then pressed it in the dark on a moist film of coloured gelatine also sensitized with bichromate. The gelatine became insoluble wherever the paper had been affected by the light, and on development with hot water a pigment picture was obtained on the paper.

SECTION VI.—THE ALBERT-TYPE.

Services of Poitevin and Tessié de Mothay—Albert-type—Methods—Use and Comparison with the Woodbury-type.

We have seen that the parts of a gelatine film containing bichromate of potash, which are exposed to the action of light, become insoluble in water, and do not swell on being moistened; at the same time they acquire the property of adhering to fatty inks. Thus if an exposed film is moistened with a wet sponge, the unaltered places only absorb the water. If, on the other hand, printer's ink be rubbed over the film, those places only which have been changed by the light retain the ink. This fact was discovered by Poitevin, the author of many valuable discoveries in photographic chemistry.

If a piece of paper be pressed on such a film of gelatine, coated with printer's ink, the ink adheres to the paper, and thus a print is obtained of the negative, under which the film had been exposed.

This peculiar mode of printing gave at first very imperfect results. The process was rendered difficult from the fragile nature of the gelatine film, the difficulty of finding the right time for exposure, the proper consistency of the printer's ink, and other obstacles. After a hundred impressions, the film of gelatine was generally so injured that it was useless. Tessié de Mothay, at Metz, obtained moderately good results with the process, but Albert, of Munich, was the first who succeeded in making it of any practical importance.

All experimenters before Albert had transferred the gelatine film to metal, to which it only adhered imperfectly. Albert poured the gelatine solution of bichromate of potash in the dark, on glass, and exposed it, after drying, glass upwards, for a short time to the light. In this way the light produced a superficial effect; the part of the gelatine adhering to the glass became insoluble, and thus firmly fixed. A negative was then placed on the film and exposed to the light. A faint greenish picture is thus produced. The exposed film is then washed in water until all the bichromate is removed, and then is suffered to dry.

To print from such a film, it is first rubbed with

PLATE VI.

To face page 237.

Print from retouched Negative.

See pp. 49, 237.

Print from untouched Negative.

EFFECT OF RETOUCHING NEGATIVES.

a sponge wetted with a dilute solution of glycerine. Those parts only of the film which are unaffected by light absorb the water. A leathern roller is inked by rolling it on a slab of marble on which printer's ink has been spread, and then lightly passed several times over the gelatine film. All places which have been affected by the light retain ink from the roller, but not so the others, and finally a well-defined picture appears on the originally almost colourless surface. As soon as it has been sufficiently inked, a piece of paper is laid upon it, and passed through rollers coated with india-rubber, the plate being laid on a sheet of the same material. In this manner the ink of the picture passes over to the paper, and produces thus an impression with all half-tones. The inking and printing can be repeated at option, and thus thousands of copies may be prepared if the plate is very firm.

These Albert-types, or "Lichtdrucke," as they have been latterly called, approach, but do not equal, the silver prints in beauty. The process is well adapted for copying pencil and chalk drawings, which are reproduced with the utmost fidelity. Herr Albert has reproduced and published Schwind's Fairy Tale of the Seven Ravens, and several cartoons of Kaulbach, by this process. It has also been used by Obernetter for printing the views taken by the photographic department of the Prussian army in the late war. The views of the Vienna exhibition, sold in the building, and by many supposed to be ordinary photographs, are Albert-types by Obernetter of Munich, who, as well as Albert, has done much in developing this process. In the annexed double picture by Obernetter we give a specimen of an Albert-type.

The brilliancy of these pictures is produced by giving them a coat of varnish.

If the prints by this process be compared with Wood-bury-types, it is seen that the latter give the shades and dark parts better, but that the white parts often appear discoloured. The Woodbury-types are also much more like photographs than the Albert-types, for the latter have a lithographic tone. It is only by coating with varnish that they are made to resemble photographs. But both methods are rather inferior to ordinary silver photography, which has never been surpassed in the uniformity of its half-tones, in the beauty of its lights, and the depth of its shadows, and which has one special advantage over both processes, and that is the ease of production. The preparation of pictures by the Albert or Woodbury processes requires a printing plate, to obtain which a more complicated plant is necessary than in the case of the ordinary silver print, and also a skilled artist. Silver printing gives good results with simple means, and even in inexperienced hands. It will therefore always be preferred for portraits, where the object is often only to produce a dozen pictures. The Woodbury and Albert processes are of great importance when the object is to produce large numbers of pictures in a short time.

SECTION VII.—ANILINE PRINTING.

Aniline Colours—Action of Chromic Acid on Aniline—Its Use in Photo-graphy—Willis's Printing Process—Its Application.

Every one knows the brilliant aniline colours—Hof-mann's violet, magenta, aniline green, and others. We are indebted for these wonderful pigments, surpassing all earlier dyes in brilliancy, depth, and reflecting power, principally to the noted chemist, Hofmann. The colours are due to the action of various oxidizing agents on aniline.

Aniline is a substance which resembles ammonia in its chemical relations, but having a different odour and a different chemical composition. The substance is obtained as a brown mass from coal tar on distillation.

If this brown fluid is treated with chlorine or nitric acid, or manganese and sulphuric acid, or arsenious acid, various shades of colour are produced. One specially interesting us is the colour formed by heating aniline with chromate of potash and sulphuric acid; the result is a peculiar violet substance—aniline violet. Chromate of potash, as already seen, plays a part in some of our photo-chemical processes, and on this substance is based the aniline printing invented by Willis.

Willis immerses a piece of paper, in the dark room, in a solution of chromate of potash and sulphuric acid. He exposes the paper under a positive picture, e.g., a drawing or copper engraving. The light shines through the white paper, and in these places the chromic acid is reduced to oxide of chromium, which does not affect aniline colours; on the other hand, the chromic acid remains unchanged under the black strokes, which keep back the light. After the exposure a very faint picture is seen of unchanged yellow chromic acid.

If this faint picture is exposed to the vapour of aniline, a brown colour is produced at the places where the yellow strokes exist, and in this manner the original faint yellow becomes well defined. To expose the prints to aniline vapour they are placed in a covered box with a piece of blotting-paper, moistened with a solution of aniline in benzine. This process produces a positive picture from a positive, and is therefore very valuable in producing faithful copies of drawings. These copies are of course reversed, as if seen in a mirror. This circumstance limits their use in many cases. We have already explained the reason of this reversal (p. 233). Copies can, however, be obtained in their proper position if the original drawing is very thin. In that case the back of the drawing is placed against the chromic acid paper, and the light is suffered to shine through it from the upper surface.

Another disadvantage of this process is that the chromic acid paper must be always freshly prepared, as it quickly spoils, and the difficulty of ascertaining the proper time for exposure. If the time is too short, unchanged chromic acid remains everywhere on the paper, and thus the whole picture is blackened by the aniline. If, again, the time is too long, the light acts gradually through the black strokes of the drawing, reduces the chromic acid, and the paper then remains entirely white in the aniline fumes, as no more chromic acid is present to form aniline colours. These circumstances limit the value of the method, and cause the *lichtpaus* process (p. 25) to be preferred to it. In England the aniline printing is practised by the inventor, who prepares copies to order.

SECTION VIII.—PHOTO-LITHOGRAPHY.

Nature of Lithography—Chromo-lithography—Zincography—Poitevin's Discovery—Photo-lithography—Its Application in multiplying Maps quickly—Its Importance in War—Difficulties—Photo-lithography with Asphalt.

By lithography is understood a process of printing from a drawing or painting on a prepared stone. Near the little Bavarian town of Solenhofen, there is a clayey, rather porous limestone, easily polished and worked. Such limestone is used for lithography. But the lithographic press differs essentially from copper-printing and book press, because the drawing on stone is neither raised nor incised. The lithographic stone forms, when ready for printing, a smooth surface; and in this the process is peculiar, differing from all other modes of printing. If a drawing is made on a lithographic stone with ink consisting of colour and a fatty substance, e.g., oil, and the stone is moistened with water, the porous stone is wetted only in those places where there is no oily colour, for oil repels water. An oily ink, such as printer's ink, is then rubbed on the stone with a leathern roller and only adheres to the previously inked spots—that is, to the drawing.

After the stone has been inked as above, if a piece of paper is pressed upon it the ink passes over to it, and a lithographic impression is obtained. The stone can be evidently used any number of times, and thus thousands of copies can be produced. This style of printing has many advantages over copper engraving. The engraving of a copper plate is a difficult matter, often requiring a labour of years, whereas the drawing on stone

R

is almost as easily made as on paper. In like manner, printing from a stone plate has fewer difficulties than that from a copper plate. Corrections are easier to make on the stone, and further, after the surface has been removed by grinding, the same stone may be used for another drawing, and so on for many years. These circumstances have brought lithography into general use : technical drawings, wine labels, circulars, visiting cards, price lists, calendars, illustrations of books, atlases, scientific pictures, and a thousand other things are produced by lithography. Of late years a development of this process, called chromo-lithography, has come into general use, and is by far the most important method of producing coloured pictures mechanically. Chromo-lithography is rather more complicated than common lithography. If it is wished to make a chromo-lithograph of a painted picture, not only one stone, but a separate stone for every colour must be prepared. For example, to prepare a picture of an object in which blue, red, and yellow tones appear, a drawing in blue ink of all the blue parts must be made on a separate stone; a second and third stone are required for the yellow and red places. Prints must be made from each stone successively on the same piece of paper in the proper position, and thus a picture in colours is obtained. If such a picture is then coated with a brilliant varnish it becomes an "oleograph." Though chromo-lithography offers great advantages for maps, ornaments, etc., and affords many excellent artistic specimens—*e.g.*, the chromo-lithographs of Hildebrandt's water-colour paintings—we must express an adverse opinion of oleography which, with a few honourable exceptions—by Prang at

Boston, Korn at Berlin, and Seitz at Hamburg—produces pictures of very small artistic value, and has done much to injure public taste.

An artist's taste and knowledge of colour are necessary for chromo-lithography, and the printers do not possess them.

Closely related to lithography is zincography, which we shall glance at here before passing to photo-lithography.

Zinc, curiously enough, has the same property as lithographic stone; for if drawings in oily chalk be made on a zinc plate, and then the whole moistened with gum water, the plate may be inked with oil colours, and thus a picture be obtained of the chalk drawing. The printing, therefore, presents results similar to those of lithography; but the preparation for zinc printing has more difficulties than lithography, so that the use of zinc for this purpose is limited.

We have given a brief survey of the principles of lithography and zincography, as far as necessary to understand what follows. Both processes resemble the Albert-type printing in many respects; in both cases the surface has the peculiarity of retaining ink in some places and repelling it in others—but the Albert-type is of recent date, while lithography has existed seventy years. When photography was invented, it deprived lithography of an important branch, that of portraits. Even in 1850 numerous lithographic portraits were made of individuals. But since the introduction of cartes de visite, portrait lithography has greatly fallen off, and is only used for cheap likenesses of distinguished persons. The lithographs from oil-paintings have also suffered through photography, which thus entered into competition with lithography. Poitevin, who allied the two by inventing photo-lithography, endeavoured to economize the labour of the lithographic draughtsman, and to replace it by the chemical action of light. He coated lithographic stones with chromate of potassium and gelatine, and exposed them under a photographic negative. The picture thus obtained was then washed and inked. All parts affected by light took the colour, and gave an impression in the press. The first attempts of the kind were very imperfect; the pictures were especially wanting in half-tones, which were lost in washing, as they are in the pigment process (see p. 233). An improvement was introduced by Asser and Osborne; they printed the negatives on paper prepared with bichromate of potash coated with gum gelatine or albumen, and then they inked this. Such paper has the peculiar property of retaining printer's ink on the exposed places only. After inking, the paper was carefully washed and then pressed on a lithographic stone. This absorbed the colour, and thus the picture was perfectly transferred to the stone, from which excellent lithographic prints could be obtained in the ordinary way. Though half-tones were thus produced, the impression fell far short of photographs in quality. The lithographic half-tone differs essentially from the photographic, which forms a homogeneous surface, while the lithographic half-tone appears as a mass of black points more or less near together. The granular structure of the stone does not allow such delicacy as may be attained by photography; therefore, photo-lithography is employed only where cheap production of many copies is of greater moment than delicacy.

Quite recently, maps of rivers and mountains have been sold by Kellner & Co., at Weimar, which have been produced by photographing plaster reliefs, and printing from the negatives then obtained by photo-lithography. The maps were produced without the aid of a draughtsman and at so low a price as to make them accessible to the poorest students.

Photo-lithography cannot be used to reproduce works of art. For this the Albert-type, with its excellent half-tones, is a formidable competitor, though photo-lithography is very useful, because a great number of impressions may be obtained from the same plate, while the number from a gelatine plate is always limited, and the prints moreover are of varying quality.

In one branch photo-lithography surpasses all other reproducing arts; that is, in producing copies of maps which have been drawn by hand. The preparation of geographical maps requires much time and care. The outlines of mountains, rivers, and countries must be executed with the greatest exactitude, corresponding to the measurement. Frequently draughtsmen and engravers are employed for the various details, and though working conscientiously, inaccuracies are unavoidable, and make corrections necessary. All this takes time and trouble. If the object is now to make an enlarged or diminished copy of an existing map, the same difficulties occur, and the diminishing is especially troublesome. The pantograph is a useful aid here, but does not exclude inattention in the draughtsman. In this respect photography is invaluable as an aid to map-making. With very great ease, enlarged or diminished copies of an original may be prepared as negatives by photography; in a few hours this is copied on stone, and within a day photo-lithography can throw off thousands of enlarged, diminished, or original sized copies.

If it were wished to make a lithographic stone drawing by hand, several days would be necessary, and it would be far less exact. No photographic printing process is as rapid as photo-lithography, and it has in consequence been much used in map-making, especially where a large number of copies are required. In the French war, the advancing troops needed, before all things, maps of the territory to be occupied. But there was not a sufficient supply of maps of France to provide whole Army Corps with. It is impossible to keep a stock of such maps, as no one can tell where a campaign will take place. Photo-lithography was here an important means of preparing thousands of copies from one original map; and thereby contributed to the successful advance of the German army, which, with these maps in hand, showed itself better acquainted with the enemy's territory than the enemy's troops themselves. The photo-lithographic establishment of the brothers Burchard, at Berlin, produced in the war of 1870-71, 500,000 maps. Plate V. is a specimen of this process.

In connection with this subject we must mention Herr Korn of Berlin, whose work belongs more to pure art. Particularly admirable are the photo-lithographic copies of the drawings made by Berg in the Japanese expedition. These are so faithfully reproduced that original and copy cannot be distinguished—the character of the originals being, of course, very favourable to photo-lithographic reproduction.

Faint drawings place difficulties in the way of photo-

REDUCTION OF
A SECTION OF THE ORDNANCE SURVEY MAP
OF ENGLAND.

PLATE V.

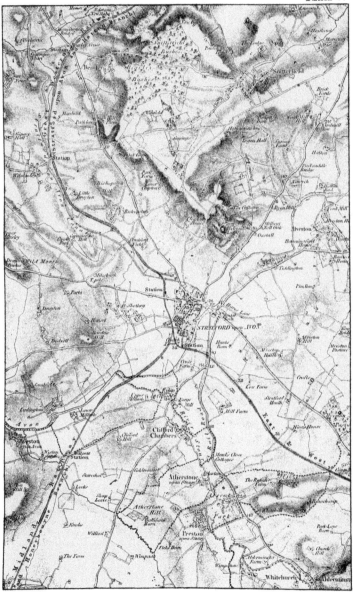

graphic reproduction, especially when they have a bluish
tint; and therefore pencil drawings are very difficult to
photograph. No perfect photo-lithograph can be produced
from an imperfect photographic negative.   Thus far,
therefore, the character of the original has a very material
influence.   Berg's drawings are executed in strong black
Indian ink, and therefore easy to reproduce.   In the
Austrian Institution of Military Geography, the map
drawings which are to be copied by photo-lithography
are so executed from the first that they make a favourable
photographic object, or, to use the technical term, come
out well. For the photographic reproduction of a drawing
brown tints, such as umber, when mixed with Indian ink,
are the best.   On the other hand, much depends on the
paper being without blemish; yellow spots, scarcely
visible to the eye, have the same effect in photography
as black ink. We knew a case in Korn's photographic
printing-house where an unblemished drawing of a map
came out as a photograph full of spots. The defect was
attributed to the chemicals, until it was found that
minute rust spots in the paper, which had got into it
during manufacture, were the cause of the defect.   In
such cases the evil can only be remedied by retouching
the negative.

The nature of photo-zincography will now be clear
to the reader : for the treatment of the zinc plate and of
the stone is the same. The negative is either copied
directly on the zinc plate, coated with gelatine and
bichromate, or a copy from the negative is prepared on
chromo-gelatine paper, and the paper is then inked and
transferred by pressure to the zinc plate.   In both cases
the prints may be taken directly from the zinc plate.

It must be remarked that, even without photography, direct mechanical copies can be made of maps, writings, etc., if the original be executed in oil colours by a transfer process. The back of the original is moistened with acidulated gum water, and then the face is inked with an oily ink which adheres only to the oily strokes of the drawing or printing. The original, thus freshly inked, is then placed on a fresh stone, or a freshly cleaned zinc plate, and pressed. The drawing passes over to the stone or the zinc, and can be easily multiplied by inking and printing. It is difficult to preserve the original, which is often much damaged by the pressure. Still more difficult is it to obtain clean lines, for the ink is often squeezed out by the pressure, and if the lines are close, as in the mountain lines of maps, they run together ; therefore the process has been more applied for copying old books, which have been reproduced page by page in this way.

It is self-evident that only reproductions of the original size can be made by this process.

We have to mention another process of photo-lithography, based on the use of asphalt. We have already described this in our first chapter as a sensitive substance, and also a process called heliography, which produces, by means of photography, copper plates and steel plates for printing. Asphalt serves also for photo-lithography. A lithographic stone is sprinkled with a solution of asphalt in ether, allowed to dry in the dark, and exposed under a negative. The asphalt becomes insoluble on the exposed places, and is retained upon treating the stone with ether or benzine. If the stone is then damped, the moisture only penetrates where no asphalt covers the stone. On rolling it after this with an inked roller, the fatty ink is

not retained by the damp places, and only adheres to the
asphalt—that is, to the picture; thus a stone giving
impressions is obtained. This method gives good results
in the hands of several practitioners, and is preferred by
many to the gelatine process, though asphalt is much less
sensitive than the prepared gelatine.

Section IX.—Encaustic Photography with Salts of Chromium.

Poitevin's Process—Action of Bichromate of Potash on Sticky Substances
    —Pictures developed by Dust—Pictures on Porcelain—Oidtmann's
    Pyro-photography—Application to the Decoration of Glass—Photo-
    graphy and Painting on Glass.

Photography has become allied to almost all the
multiplying and descriptive arts, though it was at first
looked upon as their rival. It is not supprising, therefore, that it has become a help in porcelain painting and
decoration. We have already mentioned (p. 207) the
peculiar process of changing silver pictures into gold and
platinum pictures, transferring them to porcelain, and
burning them in. This method might be called a wet
process; the same end can be obtained by a dry method,
and by the help of salts of chromium. The original
method was invented by Poitevin, and subsequently was
materially improved by Joubert in London, and Obernetter at Munich. It consists in this : a mixture of gum,
honey, and bichromate of potash is poured on glass; the
film is carefully dried in the dark, and then exposed
under a positive. This film of gum, when freshly prepared, is sticky and retains any colouring matter strewn
over it in powder, but after exposure to light the film
loses its stickiness. If this exposure takes place under a

drawing with black strokes, the film under them will retain its stickiness, but lose it beneath the white, transparent parts of the paper.

Therefore, if any colouring matter in powder be strewn over the film after exposure, it adheres where the strokes of the drawing have protected the film from the light, but not at other places, and thus a picture in powdered colour is obtained. If this coloured powder and the surface on which it is strewn are fire-proof—as glass and porcelain—the picture obtained can be burnt in, and pictures of various shades can be produced, according to the choice of the powdered colour. Pictures of this kind can be transferred from one surface to another. If a collodion film is poured upon the powdered picture and suffered to dry, and then the whole immersed in water, the collodion film with the picture can be easily separated from the surface on which it was produced, and transferred to and burnt into other surfaces— glasses, cups, etc. Thus Joubert in London has actually burnt in large pictures on glass. Obernetter at Munich, and Leth at Vienna, Leisner at Waldenburg, and Stender at Lamspringe, Greiner at Apolda, and Lafon de Camarsac at Paris have also produced encaustic pictures on porcelain in the same manner.

This process is not much employed for portraits. On the other hand, Oidtmann at Linnich,* has employed it advantageously in glass manufacture. He has copied tapestry pictures from lithographs directly on glass, and burnt them in, thereby producing cheap window ornaments. At the Vienna exhibition, there was over the

* See "Photographische Mittheilungen," Jahrg. 1869. Berlin: Oppenheim.

door of the Emperor of Germany's pavilion a rosette ten
feet in diameter, produced by Dr. Oidtmann by the above
process.   The same person has also employed the process
to produce mosaic glass pictures, similar to the mediæval
paintings on glass.   These mosaic glass pictures are
produced by cutting out coloured pieces of glass corresponding to the figures and their colours.   For example,
for a human figure, the outline of the face was cut out of
a flesh-coloured glass plate; the same thing took place
for the drapery, in glass plates corresponding to the
colours.   The lights and shades and details—for example,
nose, mouth, and eyes—were then drawn with black
fusible colour on the proper piece of glass assigned to it,
and burnt in, after which all the separate pieces of the
glass were set together with lead.   Dr. Oidtmann does,
by means of photography, what the draughtsman does in
this mosaic glass-painting.   He copies the outlines of the
face from the large-sized original lithograph, or woodcut,
on the proper piece of glass, and powders it with fusible
black colour, and thus obtains a picture which can be
burnt in, and treated in the manner described.   At the
Vienna exhibition there was a copy of " The Crucifixion"
by Dürer, produced in this manner, and composed of
150 glass pieces.   Dr. Oidtmann prepares the large-sized
original pictures by magnifying small woodcuts by photography.   (See p. 91.)   Dr. Oidtmann has also attempted
to produce pyro-lithographs, by proceeding on the
principle of chromo-lithography.   (See p. 242, Photo-lithography.)   He copied the similarly coloured parts of a
painted drawing—covering over the others—on a film
of gum and chromium, powdered this with a suitable
colour, and then copied the other colours of the original

successively in the same manner. He thus obtained a
picture, in powders of various colours, which could be
burnt in.

The Nature of the Sand-blast Process—Its Employment in Heliography
instead of Corrosive Acids.

Tilghmann, at Philadelphia, made the observation,
during his residence at the watering-place Longbranche,
that the windows exposed to sea wind became quickly
dim. He found that this was occasioned by fine sand,
which the wind drove against the window; this gave
him the idea of making ground glass by blowing sand
against it, and he succeeded perfectly. He covered a glass
surface with an iron screen, in which figures and letters
were cut; and kept this in a current of air and sand. In
a short time the exposed parts of the glass were ground,
and thus a picture of the figures produced on the glass. A
blast of a pressure of only four inches of water and a period
of ten minutes are required for this work. If the air
pressure is stronger, or steam is used to convey the sand,
at a pressure of 60 to 120 lbs. to the square inch, the
effect is wonderful. Sand blown with such power through
a narrow pipe bores deep holes into the hardest stones,
and even into glass. The process has been used to bore
stone and metal plates. Figures cut in a cast-iron screen
placed over a stone surface, can be deeply engraved in a
short time in the stone. The iron plate is also affected,
but much more slowly than the stone slab. A cast-iron
plate $\frac{3}{16}$ of an inch thick is only reduced $\frac{1}{16}$ of an inch,
in the same time that a cut 300 times deeper is made in

marble. India-rubber endures the sand stream almost as well as iron. Marble, protected by a perforated screen of india-rubber, may be cut 200 times as deep as the screen is thick without perceptibly affecting the india-rubber.

With the pressure of 100 lbs., such a sand stream can penetrate in one minute $1\frac{1}{2}$ inches deep into granite, 4 inches into marble, and 10 inches into soft sandstone.

The circumstance that soft bodies can be used as shields has led to elegant applications of this method in the industrial arts. For example, if glass be covered with lace pattern and a sand-blast be directed on it, the glass becomes ground in the meshes, and a copy of the lace is obtained on glass. In the same manner, paintings with a gum colour upon glass can be produced clear on an unpolished ground by the sand-blast. This circumstance led immediately to the application of photography. If a gelatine picture is produced on glass by the transfer process (see above), the surface of the glass at all the dark spots of the picture is protected by a layer of gelatine. If now a sand stream is allowed to operate upon it, it roughens the glass only at the uncovered places; thus a picture is formed. If the gelatine picture is a negative, the shadows are dim. A plate ground in this manner may be used for printing with printer's ink. The metal plates of Talbot's helio-graphic process may be etched by the sand-blast instead of by acids, which often eat sideways into the fine strokes, making them too wide. The sand, in conse-quence of its direction being at right angles to the plate, cannot widen the lines, and therefore the incisions may be carried to a great depth, and then the plates may be used with letter-press.

Tilghman suggests that a gelatine positive should be produced upon a cake of resin; that this should be blown upon and deeply hollowed out. A cast in plaster from this will form a mould for a type-metal cast, which could be used for printing.

These are interesting experiments, which ought in time to lead to important practical results.

SECTION XI.—THE PHOTOMETER.

In many of the above described processes, it is very important to determine the exact time for exposure. This is not easy, because the picture appears only faintly

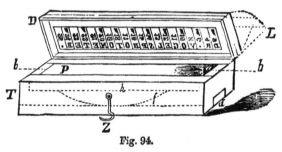

Fig. 94.

or not at all, therefore the appearance of the picture gives no safe criterion respecting its completion. This circumstance has necessitated the application of a photometer, to determine the duration of the exposure. Such photometers have been constructed by Byng and Swann in England, and by the author. The author's photometer consists of a semi-transparent paper scale *L* (Fig. 94), whose transparency diminishes from 2 to 25.

This scale is formed of layers of paper, the number of which is expressed by the figure printed on them; under this scale is exposed a strip of chromated paper; that is, paper which has been plunged into bichromate of potassium. The strip is enclosed in a little box in such wise that, when the cover D with the scale is shut down, the paper and the scale are in close contact; the light now shines through the scale and browns the paper strips lying under it. The paper is first affected under the thin transparent part of the scale, and thence to the opaque end, the rapidity depending on the strength of the light. To know how far the effect of the light has extended, figures are printed on the scale, which do not permit the light to pass; they therefore remain clear on a brown ground, and the place to which the effect of light has advanced is indicated by the figure that appears there.

To use this instrument, some experimental prints must be made first. Supposing it were desired to prepare a pigment print from a negative, the film of pigment is exposed under the negative at the same time as the photometer. After some time, lamplight is used to see how far the photometer paper is browned. The significant number is noted—photometer degree—and half the negative is covered, the other half continuing to be exposed until a higher degree of the photometer. Then the pigment picture is developed, and the degree of the photometer is determined where the favourable result has been obtained. Rarely more than one attempt has to be made; when this has determined the degree up to which the negative must be copied, the time of exposure can always be regulated with the help of the photometer. Practised hands only determine the degree with a few negatives, and easily ascertain up to what degree a fresh negative must be printed.

SECTION XII.—THE CHEMICAL ACTION OF LIGHT, AND THE
PEA-SAUSAGE.*

In the campaign of 1870, the well-known pea-sausage
was one of the most important articles of food for the
army, and was prepared daily by thousands. The fabri-
cation of the interior portion caused little difficulty, but
the obtaining so many skins created much difficulty. As
the supply fell short, a substitute was sought in vegetable
parchment. This paper, which is produced by dipping
blotting-paper in sulphuric acid for about a second, then
washing and drying, is distinguished by its skin-like
properties of resistance. It is impenetrable to water,
and difficult to tear. It is therefore used for the production of bank-notes. It was attempted to make
sausage skins of this paper, by doubling a sheet cylindri-
cally and pasting it together. No glue or gum can
however resist the effect of the boiling water in which
the sausage has to be cooked, and so the artificial sausage
skin fell asunder. Dr. Jacobson solved the problem by
producing an adhesive substance, with the help of the
chemical action of light, which could resist boiling water.
He mixed the glue intended for the sausage skin with
bichromate of potash, and exposed the glued parts to the
light. This made the glue insoluble, and now the artificial skin endured boiling water thoroughly well. The
number of sausage skins prepared in this way, by the
chemical action of light, amounted to many hundred
thousands.

* Erbswurst.

CHAPTER XVI.

PHOTOGRAPHY WITH IRON, URANIUM, AND COPPER COMPOUNDS.

Historical—Combinations of Iron—Action of Light on a Solution of
Ferric Chloride—Ferric Chloride and Paper—Blue Iron Pictures—
Iron-gold Pictures—Process with Salts of Iron—Iodine Pictures—
Combinations of Uranium—Uranium Pictures—Their Development
—Copper Pictures of Obernetter.

WE remarked further back that the number of substances
sensitive to light is much greater than appears at first
sight; and in fact, on close investigation, it would probably be found that all substances are more or less
sensitive. Even in the first period of photography, in
the year 1840, Herschel observed the sensitiveness of
salts of iron, Burnett that of salts of uranium, and
Kratochvila prepared successful daguerreotypes on copper
plates, in a manner analogous to that for silver plates.
The latter process has been energetically cultivated, but
hitherto without any important result.

It has long been known that a solution of ferric
chloride, a yellow substance composed of chlorine and iron
in ether, is bleached by light, the ferric becoming ferrous
chloride, a colourless salt which contains less chlorine.
The same change is produced in contact with paper;

S

thus, if clean paper be soaked in a solution of one part of
ferric chloride in six of water, dried in the dark, and
exposed under a negative, the yellow paper becomes
white beneath the transparent parts of the negative,
because the ferric chloride there becomes ferrous chloride.
The faint picture thus produced may be easily coloured
intensely dark by immersion in a solution of red prussiate
of potash.   This salt combines with the ferrous chloride
to produce Prussian blue, but leaves the ferric chloride
unchanged.   In this manner a blue picture is obtained.
If such a picture be plunged in a solution of gold, it becomes of a light blue colour, because the ferrous chloride
produces a precipitate of metallic gold. And, similarly, all
substances which form dark precipitates with ferrous
chloride will serve to develop such pictures.

   Another process consists in transforming the iron
pictures into iodine pictures.   A positive (for example, a
drawing) is printed on a piece of paper sensitized with
ferric chloride; the print comes out in yellow lines of unchanged ferric chloride, on a white ground.   If this is now
immersed in a solution of iodide of potassium and starch,
the ferric chloride decomposes the iodide, and sets free
iodine, which combines with the starch, forming the
black-blue iodide of starch, and thus colours the lines of
the picture.

   There are several other methods of making the colour
of the iron pictures darker.   The pictures in Prussian blue
gradually fade, for this dye is bleached by light (blue
parasols rapidly lose their colour in the light).   The
same remark applies to pictures of iodide of starch; the
gold pictures are too pale, and their preparation too
costly.

The salts of uranium present the same phenomena as the salts of iron. Uranium itself is a rare metal, the salts of which are much used as colouring materials; thus, there is a yellow oxide which, when burnt into porcelain, colours it dark green, and fuzed with glass, imparts to it a beautiful grass green (uranium glass).

Two chlorides of uranium—uranous and uranic chlorides—corresponding in their composition and properties to the ferrous and ferric chlorides, are also known. The best known salt of uranium is, however, the nitrate, which is reduced by light to uranous nitrate in the presence of organic matter (for example, paper). If a piece of paper be immersed in a solution of one part of the salt in five of water, dried in the dark, and exposed under a negative, a very faint picture is produced, which consists of uranous oxide. If the print is now immersed in a silver or gold solution, the picture becomes at once visible, since uranous oxide precipitates gold and silver as such from their solution (silver as a brown, gold a violet powder).

Uranium is, however, too scarce and too dear to be employed generally in photography.

As can be perceived, the salts of iron and uranium are analogous to the salts of chromium, by only being sensitive to light in the presence of organic bodies. In a pure state, salts of uranium and iron do not change in the light.

The sensitiveness to light of salts of copper has hitherto only been studied very imperfectly. Copper forms with chlorine a green salt, soluble in water— cupric chloride,—which is reduced to cuprous chloride in the light. Obernetter took advantage of this fact, by mixing chloride of copper and chloride of iron together, and saturating paper with them. This was exposed to light under a negative, then plunged in sulpho-cyanide of potassium, and ultimately treated with red prussiate of potash. The result produced by this somewhat complicated process was a brown picture.*

* See Vögel, " Lehrbuch der Photographie," p. 32. Berlin : Oppenheim.

CHAPTER XVII.

THE CHANGE OF GLASS UNDER THE INFLUENCE OF LIGHT.

Faraday's Observation on Manganese-glass—Change of Mirror-glass in the Light—Almost all Kinds of Glass are Sensitive to Light— Gaffield's Experiments—Disadvantages of the Change of Glass in the Light—Explanation of the Change of Manganese-glass—Action of Light on Topaz.

THE celebrated physicist Faraday made the observation that flesh-coloured glass, which is stained with manganese, became rapidly brown in the light. This fact remained for a long time isolated, until, years later, other observations of the same kind were made.

A very handsome plate of glass was exhibited in a mirror shop at Berlin. It bore the inscription "Spiegel-manufactur" in brass letters. After being exhibited for years, the business was broken up, and the mirror, on account of its beauty, was taken away by its owner, the brass letters were removed, and the plate cleaned. To the surprise of the proprietor, the letters remained plainly visible on the glass, notwithstanding all attempts to remove them. The surface was even ground away, but this did not produce any effect on the letters. It was found that the glass had become yellow throughout, and that it remained white only at the places where the opaque letters had kept off the light. The plate of glass was cut into halves. One half, with the word "Spiegel," remains in the Physical Museum of the University of Berlin.

Some very interesting observations on the action of light on glass have been recently made by Gaffield, who finds that almost all sorts of glass are affected by light, and that in many cases an exposure of only a few days suffices to produce a change. Gaffield went systematically to work in his experiments. He cut the glass to be examined into two parts, placed one in the dark and the other in the light, and compared the two after a few days. In almost all cases he remarked that the colour of the glass became deeper on exposure to light. Two kinds only of German and Belgian green window glass were unaffected. The exposed glasses recovered their original colour on being heated to redness.

This alteration of glass by light has a most injurious effect in photography. The yellow tinge, which the glass gradually assumes, absorbs a part of the chemically active light. This absorption makes itself strikingly evident, because the time of exposure for photographs must be continually lengthened.

The greatest change is produced in glass containing manganese. Peroxide of manganese, or black manganese, is frequently added to glass to bleach it. The oxygen of the peroxide converts the dark-green ferrous oxide contained in the glass into the lighter ferric oxide. In the light the converse reaction takes place. The ferric oxide is reduced to ferrous oxide, the oxygen combines again with the manganese to form black manganese, and thus the colour of the glass becomes deeper.

In many minerals the opposite effect to that in glass

is produced on exposure to light; their colour becomes paler instead of deeper.   This is the case with the Siberian topaz, which soon loses its golden-yellow colour in the light.   A splendid crystal of topaz, six inches long, belonging to the Mineralogical Museum at Berlin, has in this manner lost much of the beauty of its appearance.

CHAPTER XVIII.

PHOTOGRAPHY IN NATURAL COLOURS.

Observation of Seebeck and Herschel—Bequerel's Coloured Pictures on
Silver Plates—Researches of Nièpce—Effect of Black Colours—
Coloured Pictures on Paper of Poitevin and Zencker—Want of a
Fixing Medium for Coloured Photographs.

PHOTOGRAPHY has already achieved grand results: but
it has still one problem to solve—the production of
photographs in natural colours. There are plenty of
coloured photographs to be seen, but in such cases the
colour has been added with the paint brush; it is a
kind of *retouche*, which in most cases does not improve
the picture. By photography in natural colours, however,
we mean the production of pictures in the colours of the
original object by light alone. Numerous experiments
in this direction have been recorded. Some attempts to
prepare coloured pictures by the chemical action of light
have indeed been successful, but such pictures are rapidly
destroyed by the same agent to which they owe their
formation. No means of fixing coloured photographs has
been discovered.

The first attempts to make coloured photographs date
a long way back. Professor Seebeck of Jena, as early as

1810, found that chloride of silver, when exposed to the solar spectrum, became coloured in a corresponding manner. This observation, published in Goethe's "Farbenlehre," ii. p. 716, passed unnoticed, until, in the year 1841, after the discovery of the daguerreotype, experiments in the same direction were made by the celebrated Sir John Herschel. He took paper saturated with chloride and nitrate of silver, let a powerful solar spectrum fall upon it, and obtained immediately, like Seebeck, a coloured image of the spectrum, agreeing, however, only imperfectly with the original. Bequerel was more successful. He ascertained that the solution of nitrate of silver in Herschel's experiments had a disturbing effect, and he worked with chloride of silver alone. He employed silver plates, which he plunged in chlorine water. A white film of chloride of silver was thus formed on the plates, and, on exposure to the solar spectrum, an image was obtained, the colours of which agreed very closely with those of the spectrum. Bequerel observed that the duration of the action of the chlorine water was very important, and he preferred at a later date to chlorinate the plates by the galvanic current. For this purpose he suspended them in hydrochloric acid from the copper pole of a galvanic battery (see p. 221). The current decomposed this acid into chlorine and hydrogen. The chlorine passes to the silver plate, and forms chloride of silver. This method enables the operator to produce a film of chloride of silver of any thickness, according to the duration of the current. The brownish subchloride of silver is thus produced, and this is chiefly sensitive to coloured light. This sensitiveness is, however, not great: it suffices to fix a powerful spectrum, but it requires a very long exposure to obtain pictures in the camera obscura; and all such pictures, unfortunately, darken in the light. Bequerel found that the sensitiveness was increased by heating the plates. This observation was turned to account by his successor, Nièpce de St. Victor (the nephew of Nicophore Nièpce, see p. 9), who made numerous attempts from 1851 to 1867 to produce coloured photographs, accounts of which he communicated to the Paris Academy.

He worked, like Bequerel, with silver plates, which he chlorinated by immersion in a solution of the chlorides of iron and copper, and then heated them strongly. He thus obtained plates which appeared ten times more sensitive than Bequerel's, and allowed him to copy in the camera obscura, engravings, flowers, church windows, etc. He relates that he not only obtained the colours of objects in his pictures, but that gold and silver retained their metallic splendour, and the picture of a peacock's feather the lustre of nature.

Nièpce de St. Victor introduced a further improvement, by covering the plate of chloride of silver with a peculiar varnish, consisting of dextrine and a solution of chloride of lead. This coating made the plate still more sensitive and durable. At the Paris exhibition of 1867, Nièpce exhibited various coloured photographs, which lasted about a week in a subdued daylight (they were shown in half-closed boxes).

Among these pictures were a couple of black pictures on a white ground, which had been copied from copper-plate engravings. These excited great interest, and justly so, because in these pictures the darkest parts of the object had apparently produced a stronger effect upon the plate than the lighter or perfectly white parts. This is, of course, the exact opposite of what takes place with ordinary photographic paper, where the darker parts come out light, and vice versâ. This production of black by black can only be explained by assuming that the black is actually not black, but that it reflects ultra-violet light invisible to the eye. (See p. 58.)

Since Nièpce, who died in 1870, the only persons who have paid attention to coloured photographs are Poitevin at Paris, Dr. Zencker * at Berlin, and Simpson in London. The two former investigators have reverted to the older process, as employed by Seebeck and Herschel, i.e., they prepared pictures on paper, only the preparation of this paper was peculiar. Paper saturated with salt was made sensitive in a solution of silver, like the photographic positive paper (see p. 50), then washed to remove the solution of silver, and afterwards exposed to the light in a solution of subchloride of tin. By this means violet subchloride of silver is formed from the white chloride of silver. The subchloride of tin only operates as a reducing medium. This paper is in itself little sensitive to coloured light; but if it be treated with a solution of chromate of potash and sulphate of copper, its sensitiveness increases considerably, so that it is easy to copy with it transparent coloured pictures. Nevertheless, the colours are never so vivid as in the original, the red tones showing themselves the strongest. After printing, the pictures are washed with water, to make them less sensitive to light. In this condition they showed tolerably well in a subdued light, but no means

* Those who take a special interest in the matter are referred to Dr. Zencker's " Lehrbuch der Photochromie." Berlin, 1868.

have been found hitherto to make them perfectly durable. Hypo-sulphite of sodium (see p. 27) cannot be employed, as it destroys the colours directly. We must hope that future investigators will succeed in supplying this deficiency. The first attempts in uncoloured photographs also failed for want of a fixing medium (see p. 6), which was only discovered seventeen years later by Herschel.

CHAPTER XIX.

PHOTOGRAPHY AS A SUBJECT TO BE TAUGHT IN ART AND INDUSTRIAL SCHOOLS.

Importance of School Photography—Its Use for Technical Institutions —Photography as an Object to be taught in Art Schools and Universities.

THE previous chapters prove how manifold are the applications of photography. It has entered into art, science, industry, and life as a new kind of written language. Photography is to form what printing is to thought. Printing multiplies what is written, photography what is drawn; nay, more, it draws automatically by chemical means. No doubt a certain technical experience in the art is required, which can only be gained by practice; but it is easy to learn, and the time cannot be distant when it will be taught as an extension of drawing in all industrial schools. Years are devoted to the study of the art of drawing, of piano playing, and other things; a course of instruction lasting half a year would suffice to teach photography.

The author has been for nine years professor of photography in the Royal Industrial Academy* at

* Gewerbeakademie.

Berlin, the only technical institution in Germany where photography has as yet obtained a place in the course of study.   It is by no means the object of this institution to train professional photographers; it only admits photography so far as it is of importance for art and science.

Practical instruction is given in this institution in the positive, negative, and other processes of photography, namely, with special reference to their use for the reproduction of drawings and for taking pictures of machines and buildings.   Other institutions are still hesitating about admitting photography.   The importance of the matter is still depreciated; what is new is viewed by many as inconvenient.   We cannot avoid introducing, in connection with this, a passage from a work recently published, "Photography as a Subject to be taught in Industrial Schools," by Professor Krippendorf of Arau. The author says :—

"Industrial schools are founded for the special purpose of preparing pupils for a life in one of the branches of art or industry, and special attention is therefore paid to such arts and sciences as are likely to be most useful, and first of all to drawing and natural philosophy.

" Industrial training must not only take into consideration that these subjects form a complete system, taking the place of the classical languages as a basis of general culture, but it must also draw into its system the new inventions and discoveries of science, in order to win new standpoints to react with more advantage on practical pursuits.

"Now, amongst such subjects which have advanced the most rapidly in the last ten or twenty years, stands photography. This art is a genuine product of science, and not merely the result of a lucky change. It has the great merit of having been first conceived as an ideal, and then practically carried out. It is therefore an art full of value in itself, based on science, and one whose productions delight and are gladly viewed by all, extending the knowledge of pupils, and giving an idealizing tendency to young minds.

"There is scarcely any other subject in which it is a downright necessity to keep in view an independent observation of the result. Physics and chemistry are taught as results of successful experiments, and give no clue for the detection of the source of errors. The pupil only observes what the teacher puts before him, and both are satisfied if the law is found in the experiments. The methods of observation cannot be thus properly learnt, and yet this is specially fitted to sharpen the judgment. But if photography is admitted in the school course, we gain a subject which fixes the attention of the pupil in a way that no other can do. The study of photography depends principally on learning to avoid the sources of error, and their origin—a further argument for its introduction in such schools.

"If we turn to the outside of the art we meet with many other grounds in its favour.

"Art and science are learnt in technical schools for practical ends. A knowledge of drawing is specially demanded on entering the engineering profession. It may even be affirmed, of two equally talented and diligent pupils, the best draughtsman will take the first place. Drawing is the centre of gravity for most technical professions, and for this reason alone technical and freehand drawing ought in no case to be neglected. But photography is destined to support these technical studies, as it is also a mode of drawing. Indeed, if it be proposed to draw a complicated machine—as a weaver's wheel—in a few minutes, photography is the only means of doing this. The labour, otherwise requiring weeks, is reduced to an affair of a quarter of an hour, and is so perfectly done that all proportions are duly observed, and the projection must be correct from whatever point it is taken, if proper lenses be used.

"If we follow the biography of gifted pupils, we often trace them, aided by Government stipends, going first on distant journeys to study modern and ancient buildings, and bringing home as faithful designs of them as possible. What a severe labour this implies for the architect, amidst a foreign population, in a trying climate, who has to make faithful sketches in a short time amidst countless obstacles! On the other hand, how it is all abridged by photography! What would not the young travelling engineer give to take plans of entire manufactories which he has only a few minutes to view? What would not the highly cultivated philologist give to retain for himself and others, in a durable form, the overpowering impressions of life in the past, which he can only feel as a transient emotion, on the classical ground of Greece and Italy? It is our duty to announce publicly, that all these wishes have become a possible and a tangible reality through the progress of photography, and that the practice necessary to effect this is easily attained."

Krippendorf omits here an important point, which is, the great value of photography to those who are devoted to practical typography, whether it be lithography, book printing, copper-plate printing, printing of paper money, porcelain manufacture, or dyeing; for in all these branches the aid of photography is very important. We refer to the chapters on pyro-photography, heliography, and chromo-photography. In these branches we see photography as an auxiliary to the multiplying arts.

Though it has done great things in this connection, we see very few heliographers and photo-lithographers. The reason of this is found simply in the fact that art schools, training lithographers and copper-plate en-gravers, entirely overlook photography. It is set aside as no art at all by persons who feel themselves artists, yet to whom it would be useful. But in the before-mentioned alliance of photography with lithography and metal-plate printing, good results can only be achieved by the operator being equally skilled in both arts. The author has often witnessed the failure of heliographers, lithographers, and photographers who tried to work by combining the two arts. It is therefore necessary that the schools of art should take the matter in hand, and if so, a new subject, the Albert-type process, hitherto un-known to the lithographer (see p. 243), must soon become domiciliated in those institutions.

But a knowledge of photography is equally important for painters. The photography of oil-paintings has under-gone a great development of late years. Adverse opinions of it are indeed uttered by strict art critics, such as Thansing, just as idealistic tourists formerly ranted against railways, because travel was thereby robbed of its poetry. These people were right from their point of view, but they could not stop the introduction of rail-

T

ways; and, though travel may have become less poetic, railways have the advantage of giving an opportunity to persons of slender means of making excursions, and thereby enriching their minds with a knowledge of foreign countries and people, and of improving their health. Photography affords to persons of small income similar advantages in the province of art. Paintings, too expensive to be purchased by any save the rich, became slowly and imperfectly known to others by the expensive medium of copper-plate engravings. These engravings were also confined to a limited circle. But now photography reproduces with the rapidity of lightning, and with perfect accuracy, the latest creations of art, whilst its cheapness makes them accessible to all. The copy is not so artistic as that of the engraver, but it suffices to bring all that is new quickly to the knowledge of all, and in spite—or in consequence rather—of this, the engraving coming after still retains its value.

The negatives from oil-paintings, however, require retouching, in order to equalize the false effects produced by colours. This *retouche* may be very injurious if carried out by unskilled hands. The most suitable hand is that of the painter who painted the original. Good painters have already successfully worked at reproducing negatives from their own originals, and the impressions from plates retouched by the artists themselves must evidently have a much higher value than those emanating from other hands. This presents a new field for the artist; but it can only be worked with good results if the art pupils have become familiar in art schools with the technical routine of negative retouching, and with taking positive impressions from them.

In conclusion we add a few words on the education of the professional photographer.

We have already shown that if portrait and landscape photography are to produce really solid results, they require a knowledge of the principles of art. Bue hitherto nothing has been done to train photographers artistically. Moreover, photography can only be raised, in an artistic point of view, when art schools render a study of art possible to the photographer. The time must be at hand when all small jealousy directed against photography must fall to the ground. Experience has already found that it is not a rival but a handmaid to the well-trained artist.

Photography can be the more readily introduced in schools, as its tuition requires much less time than drawing, the results of which are often out of all proportion to the time spent in learning. Four hours a week for six months suffice to train a pupil in photography to enable him to go on by himself, even without a knowledge of chemistry.

Schools of science, as well as of art, must also attend to this subject, because photography is very useful as an aid to natural science.

This new art gives beautiful illustrations for science and art lectures by the magic lantern. The investigator can by its means give faithful original pictures of the results of his labours (see page 93). Hitherto proper apparatus was wanted; the magic lanterns sold in Germany gave imperfect images. Latterly, R. Talbot, in Berlin, has introduced American magic lanterns, which are best adapted for lectures.

The Woodbury press has been joined to the above for

the illustration of lectures; and the latest improvements in dry plate photography have had the result that dry plates, like *lichtpaus* paper, have become articles of trade, making the production of photographs much more easy. Thus one improvement combines with another to make photography what it ought to be—a universal art of writing by light!

INDEX.

THE END.

PRINTED BY WILLIAM CLOWES AND SONS, LIMITED, LONDON AND BECCLES.